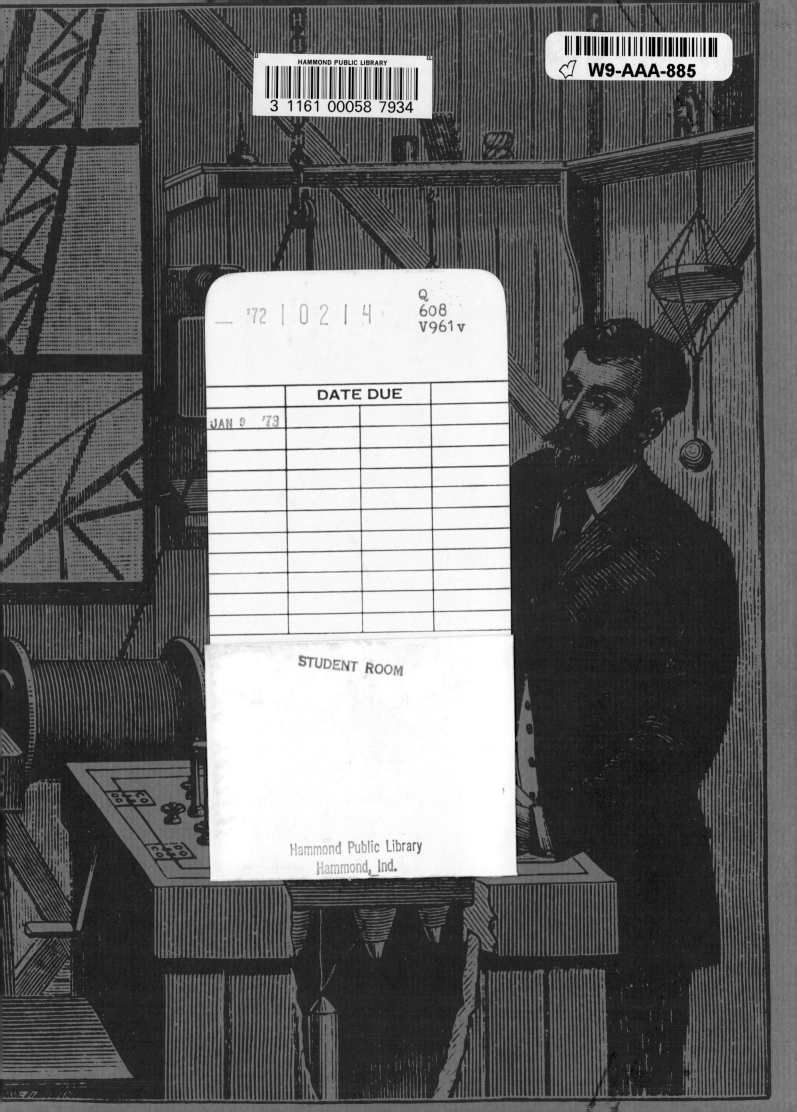

VICTORIAN INVENTIONS

translations from the Dutch by Barthold Suermondt

Leonard de Vries

VICTORIAN INVENTIONS

compiled in collaboration with
Ilonka van Amstel

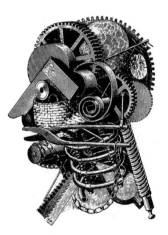

AMERICAN HERITAGE PRESS

A Division of McGraw-Hill Book Company
New York   St. Louis   San Francisco

Library of Congress Catalog Card Number: 72-38736
07-016635-8

Reprinted 1972

Designed by Ilonka van Amstel
Bookjacket by Mirja de Vries

Printed in Great Britain by
Jarrold and Sons Ltd, Norwich

INTRODUCTION

Q
608
V961 v

In a recent radio broadcast I heard somebody say: 'At the end of the last century, when trains moved slowly and few if any telephones existed . . .'—which is less than fair to our nineteenth-century forefathers since as early as 1847, in England, a train running between London and Birmingham attained an average speed of 55 m.p.h. and a top speed of 75 m.p.h. while in 1880, 50,000 Americans had a telephone, and by the turn of the century there were over three million registered telephone subscribers.

These and similar statements in the press and on radio and television repeatedly show how greatly the nineteenth century, particularly the second half, is underestimated. Nearly all the inventions we use in our everyday life date from that period. To mention but a few: the typewriter (1867), the telephone (1876), the phonograph (1877), the electric incandescent-lamp (1879) and, before the end of the century, the automobile, the electric tramcar and train, X-rays, the cinema, the wireless. Most electrical domestic appliances of today such as hot-plates, kettles, cooking-pans and hot blankets were shown at the Exhibition held in Vienna in 1883, and the electric vacuum cleaner, the washing-machine with spin-drying system and the electric dish-washer were already in existence.

What was the situation around 1865, the year in which this collection begins? In transport we find the anomaly that the world's railways, totalling some 100,000 miles in length (four times the circumference of the earth), provided fast, good and cheap travel with some very comfortable Pullman carriages, and steamships, including the luxuriously appointed, though unsuccessful, *Great Eastern* that could carry four thousand passengers, sailed the seven seas while road traffic was in many ways as primitive as two thousand years before and was dependent on the horse. It is true that since 1830 steam omnibuses had occasionally been seen, but these were heavy, cumbersome vehicles which so ruined the roads that they never became a success.

Although Siegfried Marcus, a Viennese, had built a carriage with an internal combustion engine as early as 1864, and Benz and Daimler had constantly been improving the construction of the motor-car since 1880, it was not until the eighteen nineties that the automobile was winning its place on the roads. Until 1896, its development in Britain was impeded by the ridiculous 'Red Flag Act' prescribing that every automobile should be preceded by a pedestrian carrying a red flag.

While in 1865 the railway locomotive had reached a high degree of perfection, the bicycle was still a primitive affair. Only after Starley of Coventry had, in 1885, developed his *Rover* 'safety bicycle' with wheels of equal diameter and rear wheel chain-drive and Dunlop had added his pneumatic tyres a few years later, did the bicycle become the means of conveyance for the millions.

In an attempt to make the air-balloon (invented in 1783) independent of the direction of the wind, a large number of airships were designed on paper in the nineteenth century. A few were actually built but their propulsion—by human force, steam, or electricity—was too weak to be efficient. It was not until about 1900 that the internal combustion engine provided the solution.

Shortly before 1850, Stringfellow built a model aeroplane with a wing span of 3 metres. A light-weight steam engine gave the necessary lift to the craft, which flew over a distance of 15 yards. But in the succeeding fifty years, the self-propelled, heavier-than-air flying machine, capable of carrying a man, still failed to materialise in spite of all the research, ingenuity, perseverance and money invested in it. An interesting development in nineteenth-century aviation was dis-

closed many years later by a patent investigation: it was found that in 1876 a young Frenchman, Penaud, had designed a monoplane based on sound aeronautical principles. It had many modern features such as a retractable nose-wheel, variable-pitch propeller blades, an aerodynamically streamlined fuselage and a transparent cockpit with control-lever, an artificial horizon, and even an automatic pilot. This design was at least half a century ahead of its time, but the genius who made it committed suicide at an early age. Thus, the most brilliant engineers and inventors failed in their efforts to build an aeroplane that would fly. The advent of the internal combustion engine gave a new impetus to controlled flight: this success was reserved for two modest American cycle builders, the Wright brothers, but their biplane, powered by a petrol engine, dates from 1903. Gliding in sailplanes was pioneered earlier, in 1891, by Otto Lilienthal.

About 1860, the Bessemer and Siemens-Martin processes opened the way to low-cost production of large quantities of cast steel for sheets, panels and joints which led to the erection of monumental structures such as Brooklyn Bridge (1867–1883), the Forth Bridge (1887–1890) and the Eiffel Tower (1889). It also caused a revolution in shipbuilding. The all-steel ship with its thin-walled hull offered a great deal more cargo space than the wooden ship, and by 1891 more than 80 per cent of the world's shipping tonnage—including sailing-ships—consisted of steel vessels. In 1870, 12 per cent of all ships afloat were steam-powered; by 1900 this figure had risen to 64 per cent. Ingenious devices designed to safeguard passengers against unpleasant movements of the ship were unsuccessful.

By 1865, forward-looking people were expecting great things from electricity—the miracle power. Inspired by Galvani's experiments with frog's legs, Volta in 1800 succeeded in developing an efficient source of electricity composed of an acid and two different metals. The eighteen-twenties saw the birth of the carbon-rod arc lamp and the electro-magnet, the thirties that of the first, primitive dynamos and electric motors, and of the electro-magnetic telegraph. In 1865, better electric generators and motors were available. So were transformers, but it was not until the eighteen-eighties that these machines came into general use. The impetus for this development was given by Edison's incandescent lamp (1879) which led to the erection of power stations. These, in turn, promoted the use of electric motors, for example, in the electric tramway which in the nineties gradually superseded the horse-drawn tramcar introduced around the middle of the century.

Photography, first introduced in 1839 by Daguerre, met with an enthusiastic response from the public who were all bent on having their picture taken. However, it was not before the eighteen-eighties that instantaneous photographs of moving objects could be made in rapid succession. The introduction of the celluloid roll-film by Eastman (1889) paved the way for Lumière to develop his cinematograph (1897). In the same year Edison projected the first talking picture by combining the bioscope with the phonograph, which he had invented in 1877.

As early as 1821 Wheatstone had endeavoured to transmit the human voice electrically by what he called the 'telephone'. Following the not entirely successful experiments by pioneers such as Reis (1863), Graham Bell in 1876 constructed the first telephone that really worked; it soon became popular. In 1879, London was the first city in Europe to have a telephone exchange. Two years later the telephonic transmission of operatic performances was demonstrated in Paris—stereophonically!—and in 1893 the telephone subscribers in a large part of France, and even as far as London, could make a selection from among a number of Paris opera programmes. By combining Hertz's electromagnetic wave experiments (1888) with the 'coherer'—a detector of such waves invented by Branly in 1891—and the aerial used by Popov to trace a distant thunder-storm, Marconi developed wireless telegraphy in 1896. His radio waves, Roentgen's X-rays and the discovery of radium, radio-activity and the spontaneous transmutation of some elements brought the nineteenth century to an exciting end, agog with anticipation of what the twentieth century would bring.

Victorian Inventions is composed of articles that appeared, between 1865 and 1900, in *Scientific American*, the French journal *La Nature* and the Dutch magazine *De Natuur* in which the scientific developments of the time in all countries were assessed. The original texts, condensed and in many cases translated, have been reproduced as faithfully as possible. Some items combine texts from more than one source.

As a documentary book this collection speaks for itself from a contemporary point of view, and therefore I feel that any comment of mine upon the individual inventions would be out of place. The reader can analyse and experience for himself the excitement of discoveries as they were made. However, many of the inventions described in this book never became practical reality—for economic reasons, or because they were technically impracticable, or simply ran counter to the elementary laws of nature. This may be said of the aerial bicycles, a fact which I personally regret, because as a regular cyclist, I and thousands of cyclists in Amsterdam, my home town, would have loved to use it, even at the risk of darkening the sky!

In our present-day society where teams of highly specialised scientists use the most sophisticated equipment to carry out systematic research leading to inventions and discoveries, what do we know of the lonely struggle which many nineteenth-century inventors had to endure for the sake of progress (or the hope of making a fortune). And how little are we aware of the tragic fate of many inventors who spent all their time, money and inventive spirit—often at the cost of their health—in the service of what must now be regarded as a ludicrous chimera!

It is hoped that this book will evoke the admiration of its readers for the ingenuity of our Victorian ancestors who ventured to create pneumatic railways, travelling staircases and *tapis roulants*; who were ecstatic about slot machines; who managed to have advertisements printed on the roadway by cycle tyres, or have them projected on to the clouds; who sawed ice from frozen lakes by means of steam power, and conveyed it in ships to tropical regions half-way across the globe; who established a laboratory 1,000 feet up the Eiffel Tower (see end-papers); who tried to make rain by attacking the clouds with explosives carried aloft by balloons; who presented their children with phonographic talking dolls, and could themselves be woken in the morning by a phonographic watch; who painted many miles of cloth to create huge, moving panoramas and give the man in the street the illusion of being a globe-trotter; who described television but never saw it; who yearned to be carried aloft, but never flew. . . .

And I thank the nineteenth-century inventors for the typewriter on which I am now typing this introduction, for the incandescent lamp and gramophone giving me light and music while I am doing it, and for the telephone which I am now going to use to tell my publisher that the manuscript is ready for typesetting.

Amsterdam, February 1971   LEONARD DE VRIES

CONTENTS

SCA: *Scientific American*   LN: *La Nature*   DN: *De Natuur*
Features in italic are illustrations only

TRANSPORT

THE PEDESPEED

Mercury, the messenger of the gods in ancient mythology, had winged feet. Some 3,000 years hence, some antiquarian, digging for relics among the ruins of American cities, will discover that the Yankee Mercury had his feet furnished with wheels, and that he probably made faster time than the Greek Mercury.

A few mornings since a quiet gentleman and a handsome youth walked into our sanctum, bringing with them a queer-looking package. It was no matter of surprise to us, for our eyes are familiar with nearly all the forms into which the genius of inventors can torture wood and metal. But while the elder of the two gentlemen entered into conversation with us the younger undid the package, disclosing a pair of wheels some 14 or 15 inches in diameter, to which were attached some stout hickory stirrup-like appendages, in the bottoms of which were foot pieces, shaped like the woods of common skates.

The young gentleman—who was subsequently introduced to us as the son of the inventor of this singular device—strapped on the wheels and commenced rapidly gliding about among chairs and tables with singular swiftness and gracefulness. A space being cleared he proceeded to execute, with seemingly perfect ease, the inside and outside roll, figure of eight, etc., amply demonstrating that the 'pedespeed' has all the capabilities of the skate, both in the variety and grace of the evolutions that can be performed with it.

Our engraving gives an excellent representation of this invention. Of course, no mere carpet knight accustomed to roll about on the common parlor skate, can use these at the first attempt. They require practice; but when skill is once attained, there is skating the year round. Had the pedespeed

been introduced on our rinks this winter during the long period stockholders have prayed in vain for ice, their stock would have stood higher in the market than it does at present. The pedespeed is light and strong, and it is capable of use on surfaces where the ordinary parlour skate would be useless. The inventor, a large and heavy man, informs us

he can use it constantly for two hours without fatigue. For gymnasiums, colleges and parts of the country where no ice ever occurs, it affords a delightful, healthful and graceful pastime at all seasons of the year.

When used by ladies shields may be employed to cover the top of the wheels so as to protect the dress.   [1870]

White's improved bicycle [1869]

Bicycle for two persons [1869]

THE VELOCIPEDE OR 'BONESHAKER'

Bicycle for one person [1869]

Samuels' patent hand crank velocipede [1869]

Tricycle for two persons [1883]

The 'Sociable' for three persons [1883]

The urge to move rapidly and easily from place to place is inborn in every human being. Thousands of years ago, man ceased to be content with the limited powers of his loco-motory organs and he looked longingly at the four-footed animals who moved so much faster than he, as well as at the denizens of the sea and sky. First of all he used the appropriate types of animal to enable him to travel faster and this led to the invention of a vehicle with wheels. Later he built boats propelled by oars and sails, balloons which sailed through the air and finally steam-driven ships and railway engines. Apart from their cost, such means of transport are principally designed to convey large numbers of people and are consequently almost entirely unsuitable for private use. This is the origin of the endeavour to invent a machine which can be propelled by a man, alone and unaided, and which will enable him to travel more quickly and easily over fairly long distances than would be possible on foot.

Thus as long ago as 1817 there appeared a kind of vehicle known as a 'draisine' after the name of its German inventor Drais. This consisted of two wooden wheels connected by a wooden frame, and perched upon this contraption a man could propel himself along by thrusting powerfully with his legs. Kirkpatrick Macmillan, of Dumfriesshire in Scotland, made a real bicycle in 1839. The first to be propelled without the rider's feet touching the ground, Macmillan's 'hobby-horse' had reciprocating cranks at the front, which were connected by rods to the rear wheels. In 1850, an instrument-maker called

Cycling on the water [1869]

Fischer hit upon the idea of fitting cranks and pedals to the front wheel, and in 1867 the Frenchman Michaux succeeded in improving the existing clumsy constructions by building a much lighter and faster machine which is now becoming familiar under the name of velocipede, bicycle, boneshaker or two-wheeler.

Not long ago we had occasion to read the following words:
'We should advise the constructors of the new-fangled iron horses on wheels to employ

A new American velocipede [1882]

A novel unicycle [1884]

their energies on inventing velocipedes which can be driven alternately with the hands or with the feet. Only under such conditions as these can cycling again become a beneficial form of physical movement. Moreover, those models which have but two wheels, or even only one, should be rejected absolutely. The constant need to preserve one's balance may perhaps be a source of pleasure to those who are skilled in the art of gymnastics, but the velocipede can only become an important aid for daily use if models with three and four wheels are designed and built. Generally speaking, people are not very fond of propelling themselves forward with their feet. It is a well-known fact that the vigorous movements made by the lower extremities while the rider is seated have even now frequently given rise to ailments of the lower part of the body. Ladies cannot make use of their feet in such a manner without offending against decency.'

We do not share those opinions and consider that the two-wheeler driven by the

The 'Duplex Excelsior Tricycle' [1883]

Velocipede boat in the public garden, Boston [1881]

Sectional views of velocipede boat [1881]

movements of the legs has by far the best chance of success, while the use of the three-wheeler will remain restricted to the transportation of two or three persons, or of persons with a poorly developed sense of balance. In any case, tricycles sacrifice speed entirely to the safety of the rider. We are equally in disagreement with the recently published statement that 'it must be regarded as Utopian if people think that the bicycle can ever be anything more than a means of amusement for the young.' If we are not mistaken, plans are already afoot in Great Britain to equip the country postmen with bicycles, and the day is certainly not far distant when many workers will ride by bicycle from their remote homes to the factories in the large towns. Here we show some of the latest models of bicycles and tricycles in the conviction that they will be used as 'the horses of those who cannot afford a horse.'

The velocipede has all the advantages of a horse without possessing any of its disadvantages. It costs nothing to maintain or feed. Only a slight effort is required to bring it into motion and in addition it provides a means of physical exercise which is as healthy as it is pleasant.

The 'Cooly' tricycle [1883]

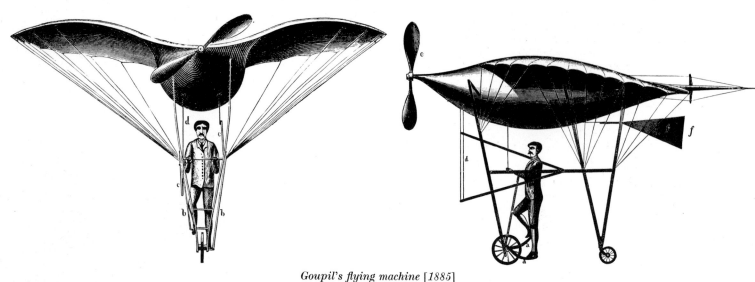

Goupil's flying machine [1885]

GOUPIL'S AEROPLANE

The aeroplane devised by Mr A. Goupil might be termed a sort of aerial velocipede.

The man, in order to obtain speed, acts at one and the same time, through the pedals $a\,a$ and the connecting rods $b\,b$, upon a wheel that moves over the ground, and through jointed arms $c\,c$ upon the helix e; and he likewise acts upon the rudder f and the tail lever by means of cords. As the apparatus obtains velocity its weight diminishes on account of the increase of the vertical reaction of the current, and, finally, it ought to ascend and maintain itself aloft solely through the motion of the helix combined with the sustaining action of the wings and regulating and directing action of the rudder. Equilibrium must be maintained through the displacement of the man's centre of gravity.

The construction of the apparatus (which is of thin strips of wood cross-braced by

AN IMPROVED SWIMMING DEVICE

We illustrate one of the most novel types of swimming apparatus permitting the user to achieve a speed of between 4 and 6 miles per hour, according to the American inventor, Mr William A. Richardson. By means of a central, longitudinal shaft, the cranking movements of hands and feet are transferred to a four-bladed propeller allowing the swimmer to proceed rapidly and easily.

———

New American swimming device [1880]

tough wood and covered with silk) is of the lightest character. The whole weighs 220 pounds. Certain persons will smile, perhaps, upon first glancing at the figures of this new aerial velocipede; and others upon reading the conditions of the apparatus' working and the hopes that are had of it, will be tempted to ask us if such apparatus have already operated—a question which we cannot answer affirmatively. However, if it is allowable to smile innocently at such claims, it is perhaps less allowable to have doubts. The rules of mechanics do not contradict the assertion that it will one day be possible for man to rise and direct himself in the air when the latter is undisturbed by storms.

When aluminium and still lighter and more powerful motors shall intervene, the solution of the problem will not have to be long awaited. But what will prove more difficult yet, after this very solution, will be the practice of the thing. It is not everything to have a sure and well-rigged ship that fulfils all the conditions of good navigation, for a crew is likewise necessary. When, then (however distant the period) it shall be felt that the end has been reached, it will be necessary to instruct the future fliers to preserve that coolness and precision of motion in the air that should contribute to secure the necessary conditions of precise manoeuvring and perfect equilibrium *Chronique Industrielle*.

A SPHERICAL, TRANSPARENT VELOCIPEDE

Imagine a hollow sphere made of some transparent, solid and not too fragile material, 5 to 7 feet in diameter, and provided with a circular opening large enough to permit a person to enter, which opening can be closed with a convex door in such a manner as not to interfere with the spherical shape of the whole.

In the centre of this sphere there is an iron shaft, with a double right-angled bend and dished ends into which there fits a metal ball. This ball presses against the wall and forms a socket-joint with it. Due to the bends in the shaft, the centre of gravity is not in the geometrical centre of the sphere so that the seat attached to the shaft will always point downwards while remaining horizontal whatever the position of the sphere may be.

It is clear that when a person sits on the seat his weight will tend to stabilise the equilibrium so that he will always remain vertical, irrespective of the direction in which the sphere is moved. The 'sphero-velocipedist' desirous of making a trip opens the door, asks somebody to hold the sphere for a moment, locks himself inside, holds the shaft with both hands to remain upright, moves his feet forward and presses them against the wall of the sphere.

After he has made a few walking movements, the sphere starts rotating in the same way as the wheel sometimes found in squirrel-cages. Naturally, the sphere moves in a forward direction and there is not the slightest danger that the person inside will lose his vertical position. If he wishes to go to the left or to the right, he need only incline his body slightly in the desired direction and the sphere will obey immediately. If he wishes to stop, he does not get up from his seat but merely presses his feet against the place where the sphere touches the ground, thus bringing to bear a friction which will presently check the sphere in its headlong career. Should he wish to move backwards, he simply reverses his steps and the sphere will respond immediately.

But that is not all. He arrives at a river—let us suppose that it is not too wide—and the sphero-velocipedist, propelling his vehicle with the utmost possible speed, rolls down the bank with sufficient momentum to bring him to the other side of the stream. For the sphere floats on the water and continues to revolve until it has reached the opposite bank.

Strange as it may seem, the project is not impossible. Objection may be raised to the fact that the sphere is hermetically sealed, thus impeding its ventilation. However, this is not a serious drawback. The sphere contains more than 140 cubic feet of air; consequently, the sphero-velocipedist will have

Bicycle for the family: the father does all the work [1896]

sufficient oxygen at his disposal for a two-hour trip.

However, there is nothing to prevent him from resting awhile every hour, and from putting his head out of the window to replenish his lungs with good, fresh air. Besides, a large number of small holes could be provided in the sphere to permit the outside air to penetrate freely to the interior. If this is done, however, it precludes the possibility of water-borne travel.

This fantasy of a new type has originated with one of the foremost manufacturers of velocipedes in France. From a mechanical point of view, his proposal is in no way impracticable. However, we do not divulge his plan in order to incite others to bring it to fruition, but merely as a scientific curiosity bearing witness to the imagination and wit of the French velocipede-maker. [1884]

A FAMILY BICYCLE

Mr von Scheidt, of Buffalo, New York, is a frequent visitor to the Niagara Falls. For these trips, of 15 to 20 miles, he uses a single bicycle on which he loads no fewer than four children. The cycle, a normal model of the make 'Eclipse', easily carries the total load of over 400 pounds. [1874]

Najork's foot motor boat [1895]

AN IMPROVED AIRSHIP

A light and strong machine for navigating the air, designed to be readily controlled by the aeronaut to give the best results in flight with the least expenditure of power has been patented by Mr John P. Holmes, of Oak Valley, Kansas, U.S.A. The horizontal frame of the machine is suspended by hanger bars or rods from an aeroplane, which is a rod frame covered on one face by a silken fabric. Towards its rear there is attached to the side bars of the horizontal frame a canvas forming a rest or support on which the aeronaut will lie, face downwards, on his breast and stomach, so that his hands may conveniently reach two transverse cranked shafts, by working one of which he can alter the incline or pitch of the aeroplane, while with the other he can rotate a propeller wheel journalled [i.e. on bearings] at the front of the

An improved airship [1889]

machine. At the rear is a rudder sail, on the sides of which lie sacks to receive the legs of the aeronaut, and allow him to guide the machine by his legs in flight. The aeroplane is arranged to be rocked up and down, and locked at any desired adjustment, for utilising wind currents and the propelling force of the wind to the best advantage.

Novel water velocipede [1895]

17

THE CYCLODROME

One of the main attractions of this winter in Paris is the Cyclodrome invented by Monsieur Guignard and recently opened at 14 Boulevard Montmartre. A bicycle placed on rollers is no longer a novelty since so many derive physical benefit from exercising on such a machine. But the cycle champions owe it to Monsieur Hurel, a Paris entrepreneur, that a genuine cycle-race—including all the thrills and excitement of such an event—can now be held in complete safety, despite the most adverse weather conditions, on his indoor race-track.

There are four contestants manning four cycles. The roller underneath the rear wheel of each of the machines is linked both to a measuring instrument which records the speed as well as the distance covered and, in addition, to a miniature cycle with a midget rider moving on the replica of a cycle-racing track. The movements of the four tiny cycles on the track correspond exactly to the distance covered by the real cycles on the rollers, thus permitting the spectators to watch the progress of the contest as closely as on a real race-course.

At the sound of a pistol-shot, the cyclists start. Pedalling ever faster and faster, they bend forward over the handlebars. The needle of the speed-indicator creeps along the scale: 25 miles per hour, 30, 32 . . . Cries of encouragement arise from the public, the contenders cast a stealthy glance at their speed-dials and intensify their efforts, they

An aerial cyclist [1888]

The Cyclodrome [1897]

18

Home trainer [1888]

A VELOCIPEDE SHOWER-BATH

At the Bicycle Exhibition recently held in Paris, a prominent English bicycle-manufacturer introduced a fascinating novelty, the 'Vélo-douche.' This device combines the morning wash with a means of keeping in training, and remaining in condition, in a truly ideal way. It is, in fact, a combination of a bicycle and a shower-bath which does not waste the driving power created by pedalling, but uses it to drive a rotating pump which forces up water from a tank. The harder one pedals, the more powerful the flow of water. A heater can also be placed below the water-tank, thus making it possible to take a hot shower-bath. In our opinion the Vélo-douche will prove a particular asset in the cycling clubs and cycling schools which have been set up in such numbers.   [1897]

pedal faster, perspiration pours from their foreheads, their speed steadily rises . . . 35 . . . 40 miles . . . One of them is slowly gaining on the others, his speed-indicator crawls towards the 45 mark and beyond . . . and to the accompaniment of loud cheers from the spectators he attains the enormous speed of 50 miles per hour!

The total distance to be covered being 30 miles, excitement still runs high among the crowd. When the winner finally crosses the finishing-line, the public burst into spontaneous applause to vent their enthusiasm over this brilliant sporting contest which, thanks to Monsieur Guignard's Cyclodrome, could be held in spite of the severe snowfall and icy roads outside!

Harper's 'Unicycle' [1894]

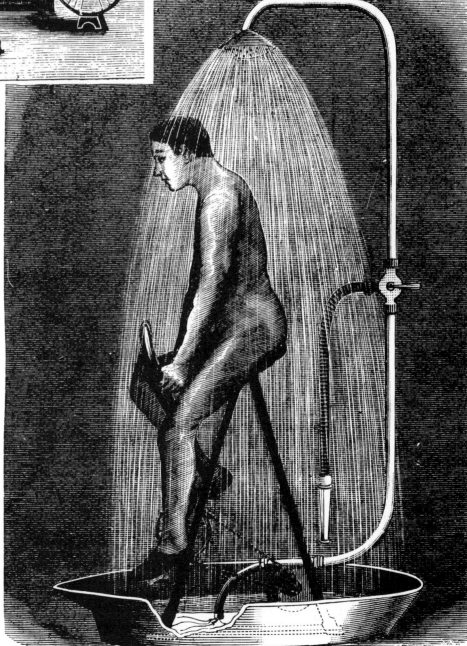

Barathon's propeller-driven lifebuoy [1895]

Cyclists in the British army [1888]

The 'Eiffel Tower' bicycle [1894]

A step, in walking and on the wheel [1895]

A PROPELLER-DRIVEN LIFEBUOY

Monsieur François Barathon of Paris has developed a swimming-buoy that can be propelled like a velocipede for the purpose of saving life at sea. The shipwrecked person sits on an inflated rubber bag placed on a curved metal plate to which a frame is attached. This frame contains a metal propeller facing rearward which is driven by the arms and a downward shaft with a horizontal propeller driven by pedalling with the feet. The occupant can also increase his speed by erecting a short mast with a sail. The device is equipped with a lamp which may attract the attention of potential life-savers after darkness has fallen.

Gauthier's monocycle [1877]

The 'Equibus' in action [1878]

The 'Equibus'—rear view [1878]

Steam hand car [1876]

STEAM-HORSE FOR STREET RAILWAYS

Mr S. R. Mathewson, a Californian, has devised a locomotive for tramways in the shape of a horse 'so as not to frighten the horses in the streets.' A 5 horsepower steam-engine is accommodated in the imitation horse's rump and to avoid the emanation of smoke, which might also scare the horses, the inventor intends to use gas for fuel. The engine is equipped with a brake capable of stopping the machine within a space of 20 feet, while travelling at a speed of almost 8 miles per hour.

A new steam carriage [1884]

Mathewson's steam horse for street railways [1876]

23

TRICYCLE AND PRINTING-PRESS COMBINED

An unusual vehicle has recently been observed in the streets of Paris: a complete, mobile printing-press! The rear wheels of the tricycle have rims to which solid rubber tyres have been secured with strong, elastic bands: on its outer circumference, each tyre carries embossed printing-types enabling all sorts of short advertisements to be composed. A tank behind the driver's seat feeds the printing-ink through a tube to rubber rollers in continuous contact with the rear wheels. Between these inking-rollers a rotating fan, driven from the wheels, blows a downward stream of air on to the street to free it from dust. In this way, the advertisement is printed on a clean background to make it legible for a prolonged period of time.

Dr Casgrain's bollee carriage for winter use [1898]

Tricycle and printing-press combined [1895]

Motor bicycle for two persons [1899]

A steam tricycle [1895]

ELECTRIC PROPULSION FOR COMMON ROADS

The illustration represents the application of a system of electrical propulsion for common roads, by means of which traffic is designed to be carried on without employing a railroad track, the steering gear being so arranged that the wagon will automatically run parallel with the line of the conductors. The wagon body to which this improvement is applied is partly supported on a castor wheel, provided with a fork, journalled in the forward end of an extension of the frame of the body. Upon the rear axle, in this case carrying the drive wheels, is mounted a spur wheel engaged by a pinion on the armature shaft of a motor secured to the main frame of the body. Above the road bed are suspended electrical conductors, supported by poles and brackets, and each wagon is provided with a trolley which rides upon a pair of the conductors, whereby connection is made between the motor and conductors, through a vertical shaft, the electrical switch

Dibble's electrically-propelled carriage [1889]

being close to the driver. The driver's seat is supported on the forward extension of the body, where he is able to guide the wagon by turning the castor wheel in one direction or another. Ordinarily the wagon will run in a line parallel with the conductors, the trolley following any deviations from a straight line, and the driver not being required to use the steering lever except when it is needful to turn aside, when the wagon can be made to run in a new line according to the position in which the driver places the steering lever. The yielding nature of the connection between the wagon and the trolley is such as to permit one wagon to turn out for another upon the road.

DOG-CART POWERED BY ELECTRICITY

In order to please his wife and their little daughter, Mr Magnus Volk, Managing Director of the Brighton Electric Railway in England, has equipped a dog-cart with a $\frac{1}{2}$ horsepower electric motor fed from six storage batteries which ensure six hours of operation. In the accompanying engraving, the inventor is seen with his wife and daughter on their way to Brighton Beach.

An electric carriage [1888]

26

Dog-cart powered by electricity [*1888*]

GREAT RACE OF AUTOMOBILE CARRIAGES

Since the early days of the present century a practical road carriage which should carry its own means of propulsion has engrossed the attention of many inventors. Today we are treated to a spectacle of an automobile carriage with four passengers which can travel 750 miles at the rate of nearly 16 miles per hour. We now present views of the prize-winners at the recent Paris–Bordeaux Race.

This race began in Paris on 11th June 1895; the course was from Paris to Bordeaux and return. The distance was about 360 miles from Paris to Bordeaux. Under the conditions of the race only four-seated carriages could compete for the first prize of 40,000 francs, or $8,000. Special prizes were also to be awarded to automatic and petroleum velocipedes. Sixty-six horseless vehicles propelled by petroleum, steam-power, or electricity and five or six petroleum bicycles competed. The preliminaries were arranged with great care, checking stations being provided to ensure the integrity of the race. Special telegraph-wires were laid along the route to transmit news of the progress of the race to Paris.

The race was witnessed by many thousands on the line of march. The first vehicle to arrive at Bordeaux was MM. Panhard and Levassor's petroleum carriage. Monsieur Levassor's time to Bordeaux was 22 hours 28 minutes over a distance of 585 kilometres (363 miles). The speed was 24·4 kilometres an hour, equivalent to about 15 miles. The

carriage of MM Panhard and Levassor met with an accident shortly after leaving Bordeaux, which delayed it over an hour, which makes the run more creditable. This carriage made the entire trip in 2 days and 53 minutes for the round trip of 1,170 kilometres (727 miles), being of the average rate of 14·9 miles per hour. We also illustrate the small carriage of MM Panhard and Levassor, which took the second prize. The carriages were constantly accompanied by bicycle riders who were soon outdistanced.

The roads in America are not good enough except in certain localities as yet to permit of a very rapid development of the automobile carriage, but their use in great cities is likely to be rapid.

Panhard and Levassor carriage: second prize [*1895*]

The Fils de Peugeot Frères carriage: first prize [*1895*]

Krieger's 'vis-à-vis' [1898]   *Jeantaud's cab* [1898]

A gas-propelled carriage [1889]

PROPOSED CITY RAILWAY

Recently prepared projects for fast city traffic in New York include the elevated railway designed by Mr R. H. Gilbert and shewn here. It calls for the erection of cast-iron arches placed 50 to 100 feet apart in the streets. The arches carry two juxtaposed tubes with a diameter of 8 to 10 feet through which carriages with passengers can be moved by compressed air. The stations situated 1,000 to 1,500 yards apart can be reached by means of lifts or elevators.

The city fathers are in favour of this project which they consider feasible and efficient, but serious objections have been raised by those who oppose what they call a 'huge bridge' in their streets.

The top of the arches may accommodate smaller-bore tubes for a very fast pneumatic post service for carrying letters and other postal articles. These are packed in cylinders for that purpose. [1872]

OPENING OF THE BROADWAY TUNNEL IN NEW YORK

The *New York Herald* says: '26 February 1870 was virtually the opening day of the first underground railway in America.' 'The *New York Times* says: 'Certainly the most novel, if not the most successful, enterprise that New York has seen for many a day is the pneumatic tunnel under Broadway. A myth, or a humbug, it has hitherto been called by everybody who had been excluded from its interior; but hereafter the incredulous public can have the opportunity of examining the undertaking and judging of its merits. Yesterday the tunnel was thrown open to the inspection of visitors for the first time and it must be said that every one of them came away surprised and gratified. Such as expected to find a dismal cavernous retreat under Broadway, opened their eyes at the elegant reception-room, the light, airy tunnel, and the general appearance of taste and comfort in all the apartments; and those who entered to pick out some scientific flaw in the project, were silenced by the completeness of the machinery, the solidity of the work and the safety of the running apparatus.'

The *Evening Mail* says: 'The problem of tunnelling Broadway has been solved. There is no mistake about it. Even as we write, a comfortable passenger car is running

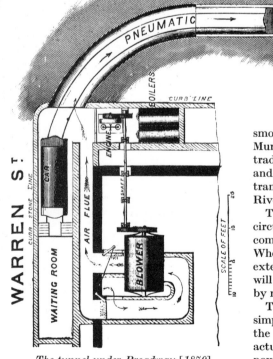

The tunnel under Broadway [1870]

smoothly and safely between Warren and Murray Streets, demonstrating, beyond contradiction, that it is only a question of time and money to give us rapid and comfortable transportation from the Battery to Harlem River.'

The passenger car used in the tunnel, is of circular form, richly upholstered, and very comfortable, with seats for eighteen persons. When the pneumatic tunnel is further extended, luxurious cars, 100 feet in length, will be used. The car is brilliantly illuminated by means of a single zircon light.

The mode of propulsion is one of the most simple things imaginable. Air is forced into the tunnel by a gigantic blowing engine, actuated by a steam-engine of 100 horse-power.

The pneumatic railway under Broadway [1870]

PNEUMATIC TRANSPORT

The growth of business and the population of New York City is wonderful. Twenty years ago there were less than 400,000 inhabitants, while today there are nearly 1,000,000 and if the same ratio of increase continues for twenty years longer there will be 3,000,000. Already the streets, spacious compared with many large cities, are overcrowded; public conveyances impede each other, and can only travel at slow pace. The carrying traffic has become so enormous, the number of men, horses and vehicles so great, that they frequently blockade the streets, move with difficulty, and of necessity their charges are high. Its costs more to carry a barrel of flour 1 mile within the streets than to transport it hither from the mills, 200 miles distant.

The city postal service, excellent in some respects, fails to afford a tithe of the assistance it is capable of rendering in the transactions of ordinary business. No person expects promptness in the delivery of parcels and letters: as for out-of-town mails, letters fail to go unless they reach the General Post Office downtown, from one to two hours prior to the departure of the car or boat.

The need of some method of relieving the streets and affording to the public more abundant, quicker and cheaper means of local communication was never more pressingly felt than at present. We are glad that a movement is being made which promises faster conveyance of passengers. We understand that the Senate Committee of the Legislature has decided to report in favour of a tunnel passenger railroad to extend from the southern extremity of the Manhattan island under Broadway, with branches under Third and Eighth Avenue, to Harlem River. Of still greater importance to the material prosperity and business convenience of the city, is the introduction of an underground method for the safe, prompt and economical conveyance of all kinds of freight, goods, parcels and the mails.

To this service the Pneumatic Dispatch is admirably adapted. This system consists of a closed tube through which air is driven or exhausted, by means of steam-power and blowers of large dimensions. Cars or trucks closely corresponding in form to the shape of the tube are employed, therein to carry freight, and these are sucked or blown along from station to station, literally with the speed of the wind. The Pneumatic Dispatch is now employed in London, with complete success. By it freight, mail-bags, etc., are transported with a velocity of 30 miles per hour, up hill and down, round the sharpest curves, with great economy. A velocity of 50 or 100 miles, or even more, per hour, may be obtained if desired, by simply burning more coal and driving the blowing machinery faster.

The Pneumatic Dispatch system is also well adapted to the propulsion of passenger cars, and for city use it is probably more economical and safer than any other known means. The superior economy of stationary engines for steady work is well known. Between pneumatic trains there can be no collisions; the same current drives them all; if one train stops on the track, no other can approach it; no engineers and firemen are required on the cars; no gas or smoke is evolved; the tunnel and cars are constantly supplied with moving fresh air; the cars run with peculiar steadiness, without any jerking at the start or stop. With an atmospheric pressure of only $2\frac{1}{2}$ ounces to the square inch on the rear end of the car, a velocity of 25 miles per hour is obtained. The use of the pneumatic passenger cars in London established these facts long ago.

The engraving is a view of the automatic letter-distributing mechanism. The packages and letters destined for different city stations are placed by the attendant in the rotary letter and parcel boxes A B C which indicate the stations to which the packages are to be delivered. The pneumatic car is divided into compartments corresponding to the boxes A B C, and when the car passes through the tube under these boxes, a pin b upon the car strikes a projection a upon the blade of each rotary box and causes it to turn upon its axis far enough to compel the contents of the box to fall into the car beneath. Each box is similarly operated by a separate pin b, and thus the contents of several boxes at the various stations on the route are successively transferred into their corresponding car compartments, without any stoppage of the car.

Application of the pneumatic dispatch to city postal service [1867]

NOVEL HYDRAULIC RAILWAY LOCOMOTIVE

A new mode of travelling has lately been invented, which the inventors claim to be applicable to any mining country where flumes exist, or which may be used wherever a stream of water of sufficient velocity of current can be enclosed for suitable distance. The device involves a carriage driven entirely by outside power; and, paradoxical as it may appear, it can travel either in the same direction as the force, or diametrically opposite thereto, while the direction of application of the power remains unchanged. In short, it is a carriage which travels up stream, impelled by no other force than that of the current.

The carriage rests on ordinary flanged wheels which traverse rails laid on the edges of the flume. On the axles are attached paddle wheels, which correspond in shape to the section of the flume and are acted upon by the current therein. It is clear that the current turning the paddles will so rotate the wheels of the vehicle, which will consequently move in a direction opposite to that of the current. When it is desired to move in the same direction as the current, the paddles are stayed stationary, and the water impels the car down stream.

Novel hydraulic railway locomotive [1877]

Mountain railway to the top of the Rigi, Switzerland [1873]

32

A SINGULAR COLLISION

For the negative of the photograph taken on the scene of this accident, a large sum of money was offered by the management of the railway company with a view to preventing the circulation of copies and so conceal the occurrence from the public. However, one copy escaped their vigilance, and we publish this unique picture of the collision between the two monsters of the iron road, an up-heaval of perhaps 60 tons of dead weight of iron. Each of the engines was pulling a goods train, and fortunately no fatalities have been reported.   [1876]

SINGLE-RAIL ELEVATED RAILWAY AND TRAIN

Mr E. S. Watson of Water Valley, Mississippi has been granted a patent for an elevated railway with only one rail. This rail may consist of normal T-sections and is supported at regular intervals by wooden poles or concrete columns. Figure 2 shows how the locomotive and coaches rest on the rail. The major part of their weight, and hence their centre of gravity, are at a lower level than the rail. Consequently, the train can never be derailed or overturn. Another advantage of this method is that road traffic can pass below the rail and no level-crossings are required.

Single-rail elevated railway and train [1884]

A PORTABLE RAILWAY

The English Army has succeeded in establishing a portable railway on several points of the Bolan Pass. This railroad is of the Decauville system, formed in sections of small steel rails, which can be put down or taken up very quickly. This ingenious railway—which has been used considerably for work on the Panama Canal and for the transportation of sugar-cane in Australia and Java—has become the indispensable means of transport in all wars. It is at present being used in Tonquin and Madagascar by the French Army, and is also being used on the Red Sea by the Italian Army.

When the Russian Government commenced the war in Turkestan, in 1882, General Skobeleff used the Decauville railroad with great success for the transportation of potable water and for all the provisions of his army. The railroad was taken up as the army moved forward, and when the

Harriman's passenger car [1890]

Russians advanced recently in Afghanistan, the little railway appeared at the advance points and was described to the English by the officers who watched the operations of the Afghans. An order for a similar apparatus was given by the English Government to Monsieur Decauville.

The object of this was, probably, that any sections of the road which might be captured from the Russians during the war could be used by the English. There was one problem which was very difficult to solve: all the material had to be carried by elephants, and they wanted a locomotive. Monsieur Decauville had the locomotive made in two parts, the larger of which weighed only 3,978 pounds, the greatest weight that an elephant can carry.

*Proposed inter-oceanic railway in Panama
—a steamer in transit [1884]*

English military railway building in India [1885]

THE STEPPED PLATFORM RAILWAY

About twenty years ago Mr Alfred Speer of Passaic, New Jersey, projected a system for city transport which consisted of what might be termed a movable pavement. He proposed to have a series of endless belts arranged side by side, but moving at different rates of speed. These belts were to be made up of a series of small platform railway cars strung together. The first line of belts was to run at a slow velocity, say 3 miles per hour, and upon this slow belt of moving pavement passengers were expected to step without difficulty. The next adjoining belt was intended to have a velocity of 6 miles per hour; but its speed, in reference to the first belt, would be only 3 miles per hour. Each separate line of belt was thus to have a different speed from the adjacent one; and thus the passenger might, by stepping from one platform to another, increase or diminish his rate of transit at will. Seats were to be placed at convenient points on the travelling platforms.

Mr Speer constructed a large working model, which operated with complete success, and was examined by thousands of people. But all the efforts of the inventor to interest capitalists to build the novel railway proved unavailing, and the patent has expired. This peculiar system of travel has lately been revived in Germany. We have received a handsome pamphlet, containing drawings, details and calculations of 'the stepped platform railway.'

One of the engravings is a diagram showing a passenger in the act of passing from one platform to another. The larger engraving shows a street scene with the railway elevated on posts. The voyager steps from one movable platform to the other, and reaches the seats. Some of these are to be covered, while others are simply open chairs. This arrangement admits of establishing a network of lines well connected in all their parts, in the business centre of a large city. Independent lines or circles built as straight lines can branch off to the outskirts of a town. The arrangement of the junction of four circuits is such that a passenger leaving one line shall get free and unhindered access to any of the three others.

The most suitable manner of working such a railway is to employ stationary engines.

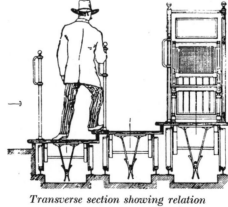

Transverse section showing relation of moving platform [1890]

which set in motion cables or chains made of suitable links, to which the sets of carriages are attached. Although it may be true that the motive power required by the running of empty trains in this system of railways is much higher than that of an urban railway, yet, without an hourly dispatch of 1,800 passengers, the motive power is but a small percentage higher than that of a railway. This result is due to the fact that the cars of the stepped platform railway can be made very light, that the weight of the driving engines has not to be moved, and that the mass once set in motion has not to be stopped and put into action again. If the traffic increases, the system shows extraordinary advantageous results over railways. With a traffic of 12,000 passengers per hour, the motive power does not even exceed one-fourth of that required by the working of a railway. To produce the same result, a railroad would be obliged to dispatch thirty trains with eight cars each in one hour. That such a traffic can occur in fact is proved by the principal lines of the London Railway Companies, which between Farringdon and Moorgate Street Stations run on four lines of rail 586 trains in one day.

The stepped platform railway will be very safe. The fall of a person passing from one platform to another would not be attended with serious results, as the difference between the speed of the two platforms is equal to the average speed of a pedestrian. Other advantages are: the number of employees can be very small; there is neither smoke nor dirt; no time-table, no late arrival; no waiting for trains.

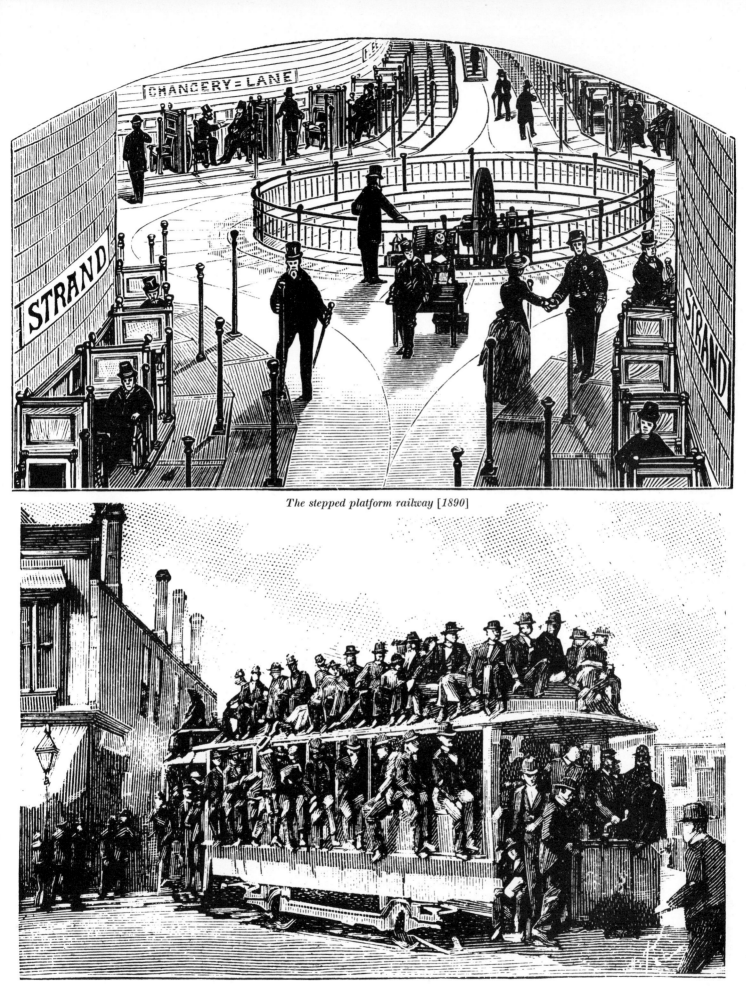

The stepped platform railway [1890]

Electric tramway in Chicago, 9 October 1893 (Chicago Day)

Kellogg's improved sleeping car [1877]

AN ELEVATED RAILWAY

The problem of fast passenger traffic in streets of restricted space can best be solved by a system of fast trains moving on rails mounted on columns and girders above two tramway tracks. The engraving shows the visualisation of this idea by the inventor, T. C. Clarke, a consulting engineer in New York.

NEW GRIP SYSTEM FOR ELECTRIC RAILWAYS

In the earliest electric railways, the two rails were used as conductors. Although this was apparently the simplest system, it was found to present many problems in practice. Perfect insulation, good electrical contact between the rails and wheels could not always be obtained because of ice, snow and mud, etc. In addition, there was the risk of electrocution both for men and animals.

A New York civil engineer, John G. Henderson, has therefore devised a system which eliminates these drawbacks. It consists of a horizontal conduit beneath the centre of the track, with its top shaped into a narrow slit. A grip shaft attached to the bottom of the carriage extending downward through

The electric railway at West End, near Berlin [1882]

The Clarke elevated and surface railroad system [1890]

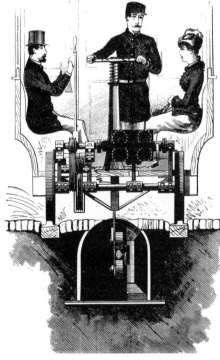

New grip system for electric railways [1884]

the slit is fitted with two insulated rollers pressing against two mating iron conductor strips on insulated mountings in the conduit.

A simplified version of the system consists of a single-strip conductor in the conduit while the rails form the other conductor.

THE ELECTRIC TRAMCAR

Until now tramcars have as a rule been drawn by horses, and more recently by steam-locomotives especially built for this purpose. There is no doubt that the horse represents the simplest means of motive power for drawing tramcars, but for large-scale transport it is just as certainly not the cheapest. A horse is an expensive tool which costs a great deal to maintain and replace. Above all, it requires a type of fuel which is far too costly to be economical. In this respect, grass, hay and bread cannot compete with coal and coke. And yet we have hitherto been compelled to use 'horse power' due to the great difficulties involved in using steam-locomotives in the streets of our cities. They terrify the coach horses, the smoke and steam they emit constitutes a public nuisance —and there is a host of similar objections.

At the Berlin Exhibition of 1879 a small railway laid by Mr Werner Siemens was the object of general attention. A few open coaches were driven along at a fairly high speed by a locomotive which outwardly appeared to serve as little more than a seat on wheels for the driver, so unobtrusive was the electric motor it contained. The current-supply came from the two rails, which were not insulated, and a centre rail which was indeed insulated.

The advantages of this Siemens's system must have been obvious to everyone present there. It eliminates the hazard of fire as far as the coaches and the roadside are concerned, for absolutely no inflammable fuel is used. For the same reason it emits no smoke or sparks. There is no steam, for there is no boiler and hence no danger of explosions.

39

Nor need a great amount of 'dead weight' be carried along; and if the vehicle is derailed its motive power is immediately cut off so that it inevitably comes to a rapid standstill.

A Siemens electric tramcar recently began running in Berlin over a distance of 1½ miles. No separate locomotive is used and the drive mechanism is so cunningly concealed that the uninitiated might well think that the vehicle is self-propelled. The voltage of the centre rail can be dangerous and it is therefore recessed in the roadway. In a circular, Mr Siemens and Mr Halske point out the great advantage of requiring a steam-engine at only one place to supply power for various tramcars and further mention that it can also be used for other purposes such as lighting.

Trials are also going on with electric tramcars in other countries, though it is questionable whether they will ever be able to run everywhere on the public highways.

Electric tram in Blackpool, England [1887]

Siemens' electric railway [1879]

POYET

Fridenberg's car cooler [1881]

THE AVITOR AIRSHIP

On 2nd July 1869 a demonstration was given in a hall in Shell Mound Lake, California, with a small model airship which its inventor Mr Frederick Mariott names the 'Avitor.' Shaped like a cigar, it is 37 feet long and the diameter in the middle is 11 feet. The 'Avitor' has two wings under each of which there is a propeller driven by a steam-engine.

The airship appeared to operate perfectly in the hall but in the open air, even in a gentle breeze, it was entirely unsatisfactory and we therefore do not expect that Mr Mariott's invention will be a practical success.

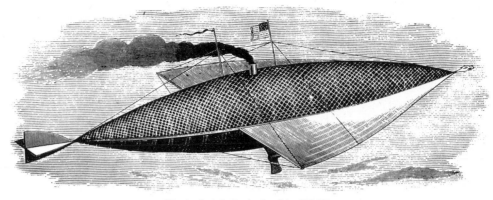

Mariott's 'Avitor' air ship [1869]

A NATURAL FLYING-MACHINE

Baltimore, 30th August 1865
Dear Editor—In recent months you have published several articles on flying-machines, undoubtedly expecting that these projects would materialise in the future. I have for years nourished an idea which, however, I never could put to the test of experiment. It is to make use of the powers of inferior animals given to man to be his servants to effect his purposes. There are many birds, for example, noted for strength of wing and endurance in flight, such as the brown eagle. Proceeding from the assumption that such birds can carry up to twenty pounds each—known as they are to carry off babies and lambs—one would require ten such eagles to convey an adult person through the air. In the accompanying drawing I have indicated how the eagles, by means of jackets fitted round their bodies, could be attached to a circular framework of hollow tubes which could carry aloft a metal basket large enough to hold a man, thus forming a natural flying-machine. An arrangement of cords passing through the hollow tubes would allow the occupant to compress or release the creatures' wings to control the altitude and, by a similar system, to regulate the direction of flight by drawing the head of the bird to one side or the other. Would not this invention lead to an extremely simple and inexpensive means of air transport?

A natural flying-machine [1865]

A NEW SYSTEM OF AERO-LOCOMOTION

The two engravings represent two forms of locomotion designed to diminish the resistance of gravity to the motion of heavy weights and that of the impact of the atmosphere against moving bodies. It is a combination of a gas of great levity with steam, hand, or other power for sustaining and driving carriages for the transmission of passengers and freight, the carriages traversing elevated roadways composed either of wire or ropes, or rigid rails, supported upon strong columns of masonry and iron combined; the rails, when of rope, to act as guys, being secured to some solid point in the earth in the manner of suspension bridges. These tracks or rails are double, one engaging with the lower surface of the wheels of the car and the other with the upper surface, the rims of the wheels being deeply grooved to ensure their retention on the track whether the weight is positive or negative.

The objects of this improvement are two: one to combine the lifting power of a gas with propelling power, to diminish the weight of a carriage in traversing levels or surfaces nearly level; and the other to provide, by means of a gas, a quick and feasible method of ascending elevations, in which case the rails or ropes are not used as tracks by which the vehicle can be impelled, but only as guides to control and direct its upward or downward movement. The aero-steamer is formed to present comparatively little resistance to the atmosphere: it is a cigar-shaped balloon

AN AERIAL STEAM-MACHINE

Joseph Kaufmann, a mechanical engineer of Glasgow, has designed a steam-powered aerial machine of which we give an illustration here. The present picture shows only two-ninths of the wings: each of these is 35 feet long and a 40 horsepower steam-engine weighing 5,250 pounds drives them up and down with a flapping movement similar to that of birds in flight. During the ascent the wings will flap almost 120 times per minute. Below the machine, suspended on a telescopic rod, there hangs a weight of 90 pounds which is designed to keep the aerial machine horizontal during its flight. Fitted with a 120 horsepower steam-engine, this could travel through the air at a speed of 56 miles per hour carrying three gondolas with a few passengers, enough fuel for ten hours and a water-supply sufficient for three hours.

Mr Kaufmann has built a model weighing 40 pounds, but this receives its steam-power from an outside source. During a demonstration the wings did indeed make movements very similar to those of birds, but when the steam pressure was increased excessively and the wings moved with undue violence, they finally broke down and the model was seriously damaged. [1869]

Fontaine's aero-steamer and self-mover [1867]

traversing the fixed guides of the stretched ropes, or the rigid rails, and moved, when on a level, by steam, wind-sails, or some other adequate means of propulsion. The frame of the structure is to be as light as comports with safety, while it is strongly braced with outside network, the lines of which pass under the compartments containing the motive power and those holding the passengers and freight. If at any time the lifting power of the gas overbalances the positive weight of the carriage, the wheels engage with the upper line of rails, and when the weight of the carriage is greater they rest on the lower line.

The self-mover is designed for ascending elevations and the vessel is filled with gas, the passengers and freight being placed in the suspended car. The ascending force of the gas is intended to raise the carriage with its load, the whole being guided by lines. In descending, the action of its gravity, assisted either by a partial exhaustion of the gas or a weight suspended to the bottom of the car, is intended to bring it gradually down to the level earth. The inventor is Dr J. A. A. Fontaine of New York City.

PROPOSED BALLOON VOYAGE TO THE NORTH POLE

We find in the London *Graphic* this engraving of an arrangement of balloons proposed by Mr Henry Coxwell as a means of crossing the Palaeocrystic Sea and so reaching the North Pole. It is believed that the three balloons connected in the manner shown would carry six men, besides 3 tons weight of gear, boat cars, stores, provisions, tents, sledges, dogs, compressed gas and ballast. The triangular framework connecting the balloons would be fitted with foot ropes, so that the occupants could go from one balloon to another in the same manner as sailors lie out upon the yards of a ship, and the balloons would be equipoised by means of bags of ballast suspended from this framework, and hauled to the required position by ropes. Trail ropes would be attached to the balloons, so as to prevent their ascent above a certain height (about 500 feet), at which elevations they would be balanced in the air, the spare ends of the ropes trailing over the ice. The boat cars would be housed-in for warmth, and telegraphic communication kept up with the ships by means of a wire uncoiled from a large wheel as the balloons move forward. The wire, being marked at every 5 miles, would also serve to keep a record of the distance traversed. Commander Cheyne proposes that the balloons should start about the end of May, on the curve of a wind circle, of known diameter, ascertained approximately by meteorological observations conducted on board the vessel, and at two observatories some 30 miles distant in opposite directions. It is estimated that, with a knowledge of the diameter of the wind circle, and the known distance from the Pole, the balloons could be landed within at least 20 miles of the long-wished-for goal. There the balloons would be securely moored; and when the necessary observations at the Pole had been carried out, a return wind would be secured for their return, the requisite full inflation having been made by means of the surplus gas taken out in a compressed condition. The returning voyagers would arrest their course to the southward on the parallel of latitude on which they left their ships, and the remainder of their journey, east or west, would be performed by means of the dogs and sledges conveyed in the balloons.

Fontaine's aero-steamer and self-mover [1867]

Proposed balloon voyage to the North Pole [1877]

Another British Polar Expedition; the start of the three balloons [1879]

Ayres' new aerial machine [1885]

ANOTHER BRITISH POLAR EXPEDITION

Our illustration depicts the three balloons as ready to start from the winter quarters of the ship during the first week in June, their destination being the North Pole. The average temperature in the early part of June is about 25 °F. The balloons are named Enterprise, Resolute and Discovery; each will be capable of lifting a ton in weight, the three carrying a sledge party intact, with stores and provisions for fifty-one days. The ascent will be made on the curve of a roughly ascertained wind circle, a continuation of which curve will carry them to the pole; but should the said curve deflect, then the required current of air can again be struck by rising to the requisite altitude, as proved by experiments that different currents of air exist according to the altitude.

About thirty hours would suffice to float our aeronauts from the ship to the Pole, should all go well. We asked Commander Cheyne how he was going to get back; his answer was cautious: 'According to circumstances,' he said, 'My first duty is to get there. When there leave it to us to get back. We have many uncertainties to deal with, and a definite programme made now might be entirely changed when the time came to carry out the journey south. Condensed gas would be taken in steel cylinders, hills would be floated over by expansion and contraction of the balloons, and in the event of any accident occurring, we always have our party with sledge, boat, stores and provisions for fifty days intact and ready for service.'

A NEW FLYING-MACHINE

Dr W. O. Ayres of New Haven in the United States of America has designed a new flying-machine so Utopian in conception that serious doubts may well be entertained with regard to its feasibility. Be that as it may, the fact that such a serious publication as the *Scientific American* has devoted space to this machine in its columns is reason enough for our decision not to deprive our readers of a short discussion of this project.

The propulsive power is derived from compressed air transported in two cylindrical vessels; this air also fills the hollow tubes in the framework of the machine. Compressed

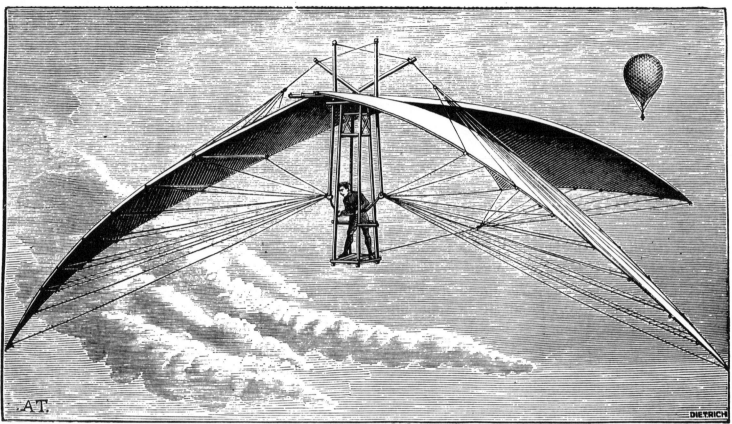

Parachute-machine of de Groof, who died on 9th July 1874

to a pressure of 200 atmospheres, the quantity of air conveyed is adequate to drive the machine for several hours.

The *Scientific American* gives further details: 'It is possible that the propellers may require to be made larger, but providing the principle is maintained, we consider that a machine such as this can do successfully what is expected of it. In order to afford support for two systems of propellers, one horizontal and one vertical, a table-like frame is required. The dimensions of this are 3 feet by 4 feet while it is supported by four legs 4 feet in height. Quarter-inch-thick steel gives the tubing all the strength needed. The rider, or aeronaut, sits upon a saddle like that of a bicycle, suspended from the top frame by steel wires.

'The four horizontal propellers serve to give the craft sufficient lifting-power. They are driven not only by the compressed air but also by the lower limbs of the rider thrusting on pedals of the type employed in bicycles. Attached to each cylinder of compressed air is a driving engine in which a paddle-wheel is brought into rotating motion by the flow of air. With his left hand the rider regulates the valve for the air-supply, while with his right arm he drives the vertically revolving propeller which thrusts the machine forward.'

DEATH BY PARACHUTE

On 9th July 1874 the Belgian de Groof fell to his death with his famous parachute. For years he had been working on an apparatus intended to emulate the flight of the birds.

A steam flying-machine [1872]

For this purpose, he constructed a device with bat-like wings. The framework was made of wood and rattan; the wings, spanning nearly 40 feet, were covered with strong, waterproof silk as was also the 20-foot-long tail. The machine was controlled by three hand-operated levers.

His first trial consisted of jumping down from a great height on to the Grand' Place in Brussels. It ended in complete failure. Fortunately, de Groof escaped unharmed. During the past summer he came to London, and there, standing in his apparatus, he was taken aloft by Mr Simmons in his balloon and released from a height of 450 feet. He glided down safely, landing in Epping Forest.

On the fatal evening of 9th July 1874, de Groof planned to descend into the River Thames with his parachute-machine. Having first taken him and his apparatus aloft to a height of 4,000 feet, the balloon descended to 1,000 feet whereupon de Groof, floating

just above St Luke's Church, released his machine from the pannier of the balloon. However, instead of bracing itself against the wind pressure, the wing frame collapsed and whirled down, dragging de Groof to a fatal fall.

Due to the suddenly diminished weight, the balloon rose quickly. Mr Simmons, the balloonist, swooned in his pannier when he

saw de Groof drop to his death, and did not regain consciousness until he was floating above Victoria Park. He landed on a railway track in Essex just in front of an approaching train, the driver of which managed to stop in the nick of time, thus avoiding a second accident. Such deplorable events as this serve to prove once more that the path of the inventor is indeed strewn with thorns.

COLE'S AERIAL VESSEL

Mr Moses S. Cole of Greytown in Nicaragua, Central America, has invented an aerial vessel which the pilot can steer in any desired direction at will. It consists of two semispheroidal balloons between which are situated the cabins for the passengers and crew, these being fitted with windows and surrounded by a circular balcony. In the middle of the lower half is a shaft containing the engine which drives the propellers; this

shaft opens into the pilot's cabin from below. A remarkable feature of the four four-bladed propellers is that their pitch can be controlled: this means that the angle which they form with their axis can be altered so that their grasp on the air can be adapted to the circumstances. The propellers shown on the left and right serve to drive the vessel forward and those which are mounted laterally lift it into the air, assisted by the balloons.

Cole's novel form of aerial vessel [1887]

'AERIAL VOYAGES'

Under the title of *Voyages Aériens* a large and finely-printed volume has been published by Messrs Louis Hachette and Co., of Paris, which deserves our particular notice. It is adorned with 117 wood engravings and six chromolithographs, designed after sketches taken by M Albert Tissandier. The book contains a minute and accurate description of all the most remarkable balloon ascents made by Mr Gaisher, by M Camille Flammarion, De Fonville, and Tissandier, for purposes of scientific observation.

We are permitted to reprint one of the illustrations. It represents an incident during Mr Gaisher's great ascent, with Mr Coxwell, at Wolverhampton on 5th September 1862, when they attained the elevation of 29,000 feet, and Mr Gaisher suffered a momentary attack of faintness, which deprived him of the power of speaking or moving, and even blinded him for an instant, though he never lost consciousness. He might have died, had not Mr Coxwell climbed to the hoop above the car, opened the valves of the balloon, and let out some of the gas, they rapidly descended to a more tolerable region.

Mr Gaisher's narratives, as furnished by himself, are translated into French, occupying the first portion of the volume, which fills above one hundred pages. *[1874]*

MAN WITH WINGS

Somebody has defined man to be a species of featherless bird. The inventor of the device illustrated herewith has aimed to supply our natural deficiencies in this respect by the provision of wings and tail, attached and operated as indicated. We hardly think he will be able to compete with the swallows in this harness, and would advise him to start from some low point at first, so that, if he should fall down, it will not hurt him much. However, we may say that the principle of calling into play the strong muscles of the thighs to aid the arms in the movement of wings is taking advantage of the greatest power the human body can exert, and in this the device is an improvement upon some other attempts. The method of connecting the rope to the various parts of the wing, is also such as gives least strain to the various parts. The machine is the invention of W. P. Quimby of Wilmington. *[1871]*

TATIN'S AEROPLANE

Although no one has as yet succeeded in rising into the air with a mechanically propelled craft, while on the other hand many a reckless fellow has had to pay with his life for attempting to descend by such means from a balloon floating at great altitude, there are nevertheless still many who cherish the belief that man will one day succeed in making wings for himself which will bear him aloft like a bird. Shall man at last succeed in doing this with the machine known as the 'aeroplane'?

The aeroplane is based on the principle that a flat, elongated surface driven rapidly forward by propellers is supported in the air. Pénaud was the first to achieve good results with aeroplanes. He used strands of elastic twisted tightly together to provide motive power for his amazingly simple constructions. Unfortunately, this ingenious inventor made only small types of aeroplanes and shortly after his successful experiments he was taken from us by a fatal illness.

'While Pénaud adhered explicitly to the use of the aeroplane as the best means of achieving practical results,' writes Victor Tatin, 'we ourselves were still busy creating machines based on the imitation of the flight of birds. Finally, we opened our eyes and set off on the road which we have never departed from since that time and which was not long in bringing success. We built a small aeroplane with a surface area of approximately $7\frac{1}{2}$ square feet and driven by two propellers rotating in opposing directions. The engine, which developed a motive power of about 19 ft-lb/sec. and weighed only 10 ounces, was very much like a small steam-engine, the boiler being replaced by a vessel containing 0·24 cubic feet of air compressed to 7 atmospheres and weighing $1\frac{1}{2}$ pounds. The total weight of the machine on rollers was about $3\frac{3}{4}$ pounds.'

In 1879 trials were carried out with this machine in the military establishment of Chalais-Meudon. The aeroplane, attached by a string to a round wooden floor, travelled round a pole and was able to rise from the ground. On one occasion it even flew above the head of a spectator, as shewn in the engraving. The aeroplane leaves the ground at a speed of 26 feet per second. These trials were of short duration and this is only a modest start. But if an airborne machine such as this were made in full size and could work at a speed of, say, 50 feet per second, amazing progress would be made and we could say that the problem of flight had been solved. New engines will be the subject of research which will not be long in yielding fruit, and mankind will finally be in possession of the most powerful tool he has ever devised: the aeroplane.

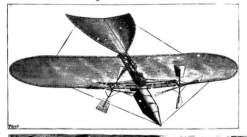

The balloon ascent of MM Sivel and Croce-Spinelli, who carried with them a considerable quantity of oxygen and made observations at heights of up to 21,000 feet [1874]

Experiments with Tatin's aeroplane at Chalais-Meudon [1879]

PETERSEN'S AERIAL WAR VESSEL

Captain Carl W. Petersen who already has 78 patents for improvements in airships to his name with, as he claims, 250 more to come, has displayed at 231 Broadway, New York City, a model of an airship of rigid construction accommodating a series of balloons filled with hydrogen gas, while an oblong nacelle is suspended underneath. Propelled by air-screws driven by electric or gas engines, this 'train' of balloons is claimed to be capable of attaining a speed of from 20 to 80 miles per hour. The vessel is extremely manœuvrable and may be used for dropping

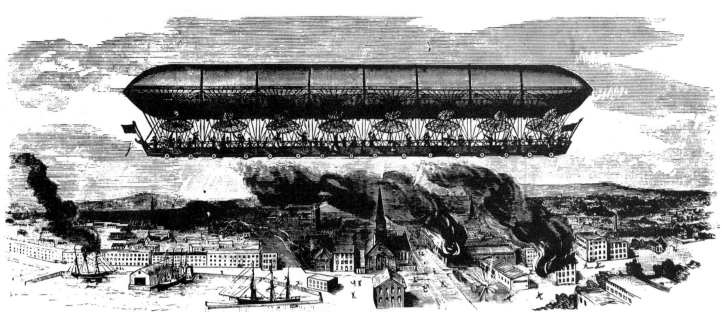

Peterson's aerial war vessel [1885]

bombs which may cause explosions and fires on enemy territory. Involving a capital of $100,000, a company has been formed to carry out the design of Captain Petersen in practice.

A NEW AERIAL APPARATUS

The offices of patent agents are swamped by inventions in the field of aerial navigation, a fact which indicates the extent of general interest in air travel. However, the results so far achieved are far from brilliant.

An interesting combination involving a giant kite and three small balloons has been invented by David Thayer, a Bostonian. It employs the—often formidable—forces of the wind in a quite unusual way. The kite is kept aloft by the balloons and, in much the same way as a folded piece of paper can be made to ascend along the kite-cord, a few men may rise in a pannier along the cables assisted by the sail hoisted above the pannier. As shewn in the drawing, a raft or boat may also be taken in tow or, on land, the kite can pull a large wagon provided with wheels and brake, or a sledge on ice and snow. This may become a practical and inexpensive method of conveying merchandise. A variation of the foregoing is the aerial train shewn in the other engraving. Also designed by David Thayer, it moves on four wheels along two wire cables strung on poles or pylons. The driving-power is supplied by an electric motor receiving its electricity from a steam-engine-cum-generator on the ground through the two cables. The train can also move under its own power provided by a steam-engine so that it may sail through the air independently of the cables which serve as rails. The advantage of such an aerial train lies in the cost of the permanent way, which is considerably lower than that of a normal railway, while the problem of crossing rivers can now be solved quickly and efficiently.

Thayer's new aerial apparatus [1890]

Thayer's dirigible balloons [1885]

Cross section of the pannier of the captive balloon in Paris [1878]

Captive balloon in Paris [1878]

Captive balloon for 40 passengers, pulled down by steam power, used in Paris 1878–1879

THE DIRIGIBLE AIRSHIP OF THE TISSANDIER BROTHERS

Gaston Tissandier, the French physicist, is the first man to apply electricity to give a balloon the power of self-propulsion and hence make it dirigible.

'In order to form some idea of the results which could be obtained,' says Tissandier, 'I first performed tests on a small scale; I arranged for an oblong balloon to be made, 12 feet long and 4 feet in diameter in the centre, with a cubic capacity of almost 77 cubic feet. Filled with pure hydrogen gas, it has a lifting-power of 2 kilogrammes. A small dynamo-electric machine of the Siemens type has been made for this balloon, weighing 220 grammes; it drives an extremely light double-bladed propeller 24 inches in diameter. With two accumulators, each weighing 1 pound, the balloon attained a speed of 7 feet per second and maintained this for ten minutes. This is one and a half times the speed of a pedestrian. The manner in which this small test balloon behaves in still air, and the speed imparted to it by the air-screw, may from

now on be regarded as offering encouraging results for airship navigation.'

Since that time, Gaston Tissandier and his brother Albert, an architect, to whom we are indebted for one of the accompanying engravings, have devoted all their energies to performing large-scale tests in their own aerostatic laboratory in Auteuil near Paris. For that purpose, they have had an oblong balloon made. The balloon has tapered ends, is 92 feet long, and has a maximum diameter of 30 feet and a cubic capacity of more than 35,000 cubic feet. Together with its valves it weighs 375 pounds. The balloon is enveloped in a network of tapes with flexible struts made of walnut and bamboo, fastened together with silk ribbons. The envelope ends in twenty ropes to which is attached the bamboo pannier with its wicker-work bottom, the two being united with rope and copper wire. The pannier contains a dynamo-machine weighing 120 pounds—this is the Siemens motor—as well as four ebonite

receptacles containing a total of twenty-four elements of bichromate of potash with carbon and zinc plates, as well as four ebonite pails filled with 100 parts water, 16 parts bichromate of potash and 37 parts sulphuric acid, all by weight. When the pail is hoisted up with a rope and tackle, 7 gallons of liquid flow into the elements which then start functioning; they can be stopped by lowering the pail again.

The electrical installation, weighing more than 550 pounds, delivers a working-power of 725 ft-lb/sec. and imparts a speed of over 180 revolutions per minute to the 15-pound propeller which is made of iron and fir-wood.

The apparatus for preparing 35,000 cubic feet of gas was completed by the end of September 1883 and the balloon lay spread-eagled on the ground under a canopy, ready to be filled as soon as the weather conditions were favourable.

Gaston Tissandier writes as follows: 'The 7th October was a beautiful autumn day

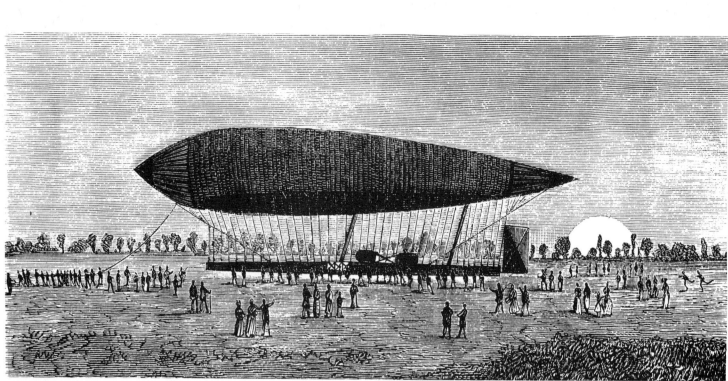

Tissandier's new dirigible airship after its landing near Chalais-Meudon, France [1884]

Experiments with Tissandier's model airship, propelled by electricity [1881]

The pannier of Tissandier's dirigible airship [1883]

Tissander's dirigible airship at its first trial, 8th October 1883

with scarcely a breath of wind. It was decided to perform the test the next day. The filling of the balloon started at eight o'clock in the morning, and it continued until half past two in the afternoon. When the balloon was completely full, the four ebonite trays, each containing 7 gallons of potassium bichromate, were placed in the pannier. Having taken in ballast, my brother and I rose into the air at twenty minutes past three. A few minutes after our departure, when we had reached an altitude of 1,200 to 1,600 feet, I put the battery into operation. A commutator filled with mercury enabled us to engage twelve, eighteen, or twenty-four of the elements so that we could give four different speeds to the propeller. With twelve elements, the airship's natural speed was insufficient, but when we were above the Bois de Boulogne and switched our motor to its highest speed, on all twenty-four elements, the balloon's inherent motion was clearly noticeable and we felt a pleasant coolness.

'When the airship was facing a headwind, it resisted the air-stream and hovered motionless over the ground. Unfortunately, it did not stay in this favourable position for long, but was suddenly caught in a rotating movement which our rudder was unable to control. When we tried to cut right across the wind, the rudder flapped like a sail and the rotations were repeated with renewed force. We then turned the motor off, and when the airship was again moved forward by the wind we restarted it. The movement of the airship then became more rapid—at 4.35 p.m. we landed safely on a plain near Croissy-sur-Seine. We left the balloon inflated overnight, and the next day found that it had lost not one whit of the gas it contained. Painters and photographers were able to depict our craft surrounded by a large and interested crowd.

56

a few defects and means are now being sought to remedy these. That the Tissandier brothers have succeeded in finding the means and that they can wing through the air at will in their craft, albeit only in fair weather, is one of the important facts which we hope to communicate to our readers in the course of this year.

Optical phenomenon, observed from a balloon on 16th February 1873

'In the daytime, we would have taken off again had not the solution of potassium bichromate crystallised so that the battery, although far from being exhausted, failed to operate. We arranged for the balloon in its captive state to be conveyed to the bank of the River Seine where we regretfully had to deflate it, thus losing within a few minutes the gas which we had so painstakingly prepared. All the material was salvaged and returned to our laboratory without suffering the slightest damage.'

This first experiment has not proved entirely satisfactory. It has brought to light

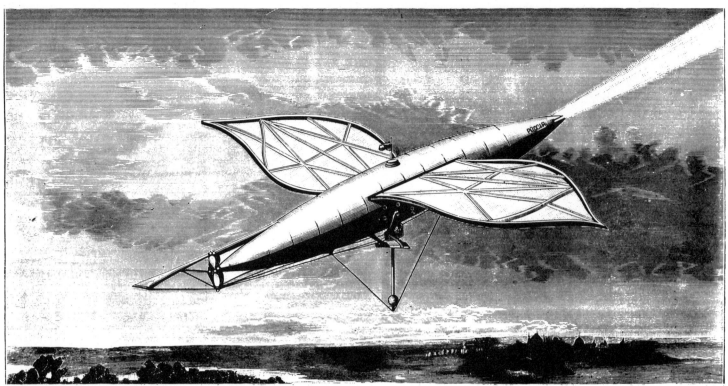

Professor Baranowski's new steam flying machine [1883]

A STEAM-DRIVEN FLYING-MACHINE

Professor Baranowski has designed a steam-driven flying-machine a model of which has flown with great success in St Petersburg, according to a report in the French *Revue Militaire*. The machine consists of a large cylinder, shaped like a gigantic bird. Inside this cylinder is an extremely powerful steam-engine which moves the wings up and down, while simultaneously driving the air-screws—one at the tail and two on the sides below the wings. The oar which can be seen to the left of the tail serves as a rudder. This has been omitted on the right-hand side for the sake of clarity. What may be described as the beak of the bird is arranged in such a way that air can penetrate to the interior to permit the crew to breathe and enable the fuel to burn. As the huge craft moves through the air, the escaping smoke and steam will cause it to look like a comet with a luminous tail. A weight suspended below the flying-machine keeps the whole construction correctly in balance.

THE PENNINGTON AIRSHIP

Despite the fact that new designs for airships and flying-machines are continually emerging, there are few which justify any hope that

On 30th November 1870, M Prince met with his death when his balloon, with which he had escaped from the besieged French capital, went down in the sea

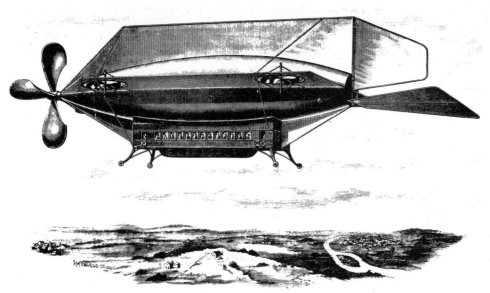

Stark's motor for flying machines [1893]

The Pennington air ship [1891]

they will satisfy the high expectations placed in them. Our picture shows an airship designed by the American E. J. Pennington with which it is intended to open up a passenger service betewen Chicago and New York. Combining the features of an airship and a flying-machine, the new craft is propelled by a large vertical air-screw, in addition to four horizontal propelling-wheels mounted on the wings for the ascent and descent. The propellers are driven by an engine in which gas is employed for fuel. Both the gas and the engine are contained in the cigar-shaped buoyancy-chamber. The cabin is suspended underneath this chamber and offers seating accommodation for some twenty to thirty passengers. Wide windows afford a splendid view of the scenery. Apart from the sail placed on top of the buoyancy-chamber to help drive the ship forward and the tail-rudder used for altering course, the entire structure is made of aluminium. Regrettably for Mr Pennington, we cannot help feeling that the chances of his project being implemented in practice are decidedly remote.

LIGHT, SIMPLE MOTOR

Mr Theodore A. Stark of Ottawa, Illinois, has designed a motor which can be set in motion with the hands and feet, thus enabling every muscle in the body to be used to supply the driving power for, say, a flying-machine. The flyer lies in a framework of tubes, his feet hook into movable rods, his arms move a second set of rods and all these movements are transmitted to an endless belt which runs over two pulleys. The motive power is then transmitted from there to the propellers.

CAMPBELL'S AIRSHIP LOST AT SEA

Aviation deplores yet another fatal accident! The unfortunate air traveller is E. D. Hogan, the Canadian aeronaut who, having made

balloon flights since the early age of sixteen, had more than 200 ascents to his credit. Hogan vanished without leaving any trace during a flight in the airship built by Campbell.

As may be seen from the drawing the airship, built at a cost of about $3,000, is

Car of Campbell's air ship [1889]

propelled by hand. Two air-screws situated above the rudder, shaped like that of a boat, serve to thrust the craft forward. Beneath the circular car, there is a multi-bladed propelling-wheel which controls the ship during ascent and descent. Built mainly from timber and wickerwork, the entire construction, suspended from the cigar-shaped balloon by means of a netting, is extremely light. Although nothing definite is known about the accident, it is most likely that this structure broke away and dropped into the water with the hapless aeronaut who thus met with his death.

———

To the North Pole by balloon [1890]

REACHING THE NORTH POLE BY BALLOON

The problem of whether there is land, ice, or an open Polar sea at the North Pole continues to be the object of heated discussions. To solve the problem, an attempt is soon to be made by MM Besançon and Hermite, a young aeronaut and a Parisian astronomer, to reach the North Pole by balloon. Having a diameter of 100 feet, it will carry up to 17·5 tons, enabling five men to take part in the expedition. The sturdy wooden car will, in addition, hold a number of husky dogs to draw the sledges, and many scientific instruments. As shewn in the drawing, it contains comfortable beds and seats, a table, a well-equipped darkroom for developing aerial photographs as well as two pens for the dogs. A rope-ladder gives access to the observation-post which affords an unimpeded view in all directions. Beneath the living-room floor there are storage compartments for food. A boat is attached to the outside of the car.

Exterior of the car [1890]

View of the interior of the car [1890]

Otto Lilienthal and his flying apparatus in two typical positions [1894]

FLYING EXPERIMENTS OF OTTO LILIENTHAL

Reports have been received from German sources about flying experiments which deserve our attention if only because they have been undertaken by a man who has made a serious study of the science of aviation. We refer of course to Otto Lilienthal who has already published a book entitled *Der Vogelflug als Grundlage der Fliegekunst* or *The Flight of Birds as a Basis for the Art of Aviation*. Mr Lilienthal wishes to apply his theory in practice and has indeed succeeded in constructing a machine in which he hurls himself from a height, remains in the air for a time, and then gradually descends to earth.

His machine consists of a framework of thin osier rods covered with fine linen and fixed securely to his shoulders. It takes the shape of two slightly concave wings with a raised tailpiece at the rear. A pair of rudders are fitted to permit the intrepid aviator to steer his course as he falls through space. The inventor has done everything possible to leave the body free; both arms rest in channel-shaped pieces padded with cushions. After a series of experiments in which Mr Lilienthal launched himself in his machine from a tower on a hilltop near Berlin, he has now transferred the scene of his experiments to a 200-foot-high-hill in the Rhinow Mountains between Tathenow and Neustadt which appeared to be particularly suitable for aerial experiments.

'When I unfolded my airborne craft here for the first time,' said Lilienthal, 'I was overcome by anxiety at the thought that I was to descend from this height into the wide expanse of landscape which stretched out far beneath me. But the first cautious attempts at diving soon restored me to the consciousness of safety, for I took off much more gently from here than from the tower which was my former launching-point.'

His manner of flight and the sensations he experienced are described by Lilienthal as follows:

'With folded wings you run against the wind and off the mountains, at the appropriate moment turning the bearing surface of the wings slightly upwards so that it is almost horizontal. Now, hovering in the wind, you try to put the apparatus into such a position in relation to the centre of gravity that it shoots rapidly away and drops as little as possible. The essential thing is the proper regulation of the centre of gravity; he who will fly must be just as much the master of this as a cyclist is of his balance. Obviously, when one is in the air there is not much time to ponder about whether the position of the wings is correct; their adjustment is entirely a matter of practice and experience. After a few leaps one gradually begins to feel that one is master of the situation; a feeling of safety replaces the initial fear. Hovering in the air you no longer lose either calmness or self-possession, while the indescribable beauty and gentle sensation of gliding along over the expanse of sunlit mountain-slopes serves merely to increase one's ardour on each occasion. It is not long before it is all one to the aviator whether he is soaring along 6 or 40 feet above the ground; he feels with what certainty he is borne along by the air, even when the tiny people down below are peering anxiously upward towards him. He travels over deep chasms and soars for several hundred yards through the air without the slightest danger, parrying the wind successfully at every moment.'

Lilienthal further reports that he has been successful in steering his flying-machine, and that during a high, long-distance flight he succeeded in making a turn of 180 degrees so that he ended up flying in the opposite direction.

It is our belief that gliding through the air might well become a sport some day, comparable, say, with cycling. Pessimists assert that man will never learn to fly. On the other hand, there are many credulous people who think that this problem has already been solved. Among these are the Russians who claim that the art of flying was invented by the Russian scientist Tchernov after he had pondered on the problem for thirty years. . . .

LILIENTHAL IN FATAL ACCIDENT

On 9th August 1896, Otto Lilienthal, whose flying experiments were recently the subject of a report by us, crashed to earth from a height of 50 feet during tests with a new type of steering device. He broke his spine and died the following day. His last words were reported to be, 'Sacrifices must be made.'

TRIAL OF MAXIM'S STEAM FLYING-MACHINE

On Tuesday, 31st July 1894, for the first time in the history of the world, a flying-machine actually left the ground, fully equipped with engines, boiler, fuel, water and a crew of three persons. Its inventor, Mr Hiram Maxim, had the proud consciousness of feeling that he had accomplished a feat which scores of able mechanics had stated to be impossible. Unfortunately, he had scarcely time to realise his triumph before fate, which so persistently dogs the footsteps of inventors, interposed to dash his hopes. The very

Trial of Maxim's flying machine [1894]

precautions which had been adopted to prevent accidents proved fatal to the machine, and in a moment it lay stretched on the ground, like a wounded bird with torn plumage and broken wings. Its very success was the cause of its failure, for not only did it rise, but it tore itself out of the guides placed to limit its flight, and for one short moment it was free. But the wreck of the timber rails became entangled with the sails, and brought it down at once. The machine fell on the soft sward, embedding its wheels deeply in the grass, and testifying beyond contradiction, that it had fallen and not run to its position. If it had not been in actual flight, the small flanged wheels would have cut deep tracks in the yielding earth.

The Maxim flying-machine is a large braced structure formed of steel tubes and wires, and is exceedingly stiff for its weight, which is about 8,000 pounds, including men and stores. At its lower part it carries a deck, on which the crew stand, where the boiler, steering wheel and reservoirs of water and gasoline are also mounted. At a height of some 10 feet above the deck come the engines, each of which drives a screw propeller of 17 feet 10 inches diameter and 16 feet pitch, working in air. Above the propellers is the great aeroplane. Smaller aeroplanes project out, like wings, at the sides, the extreme width being 125 feet and the length 104 feet. There are five pairs of wings, but the intermediate three pairs are not always used, and at the time of the accident they were not in place. Forward and aft of the great plane are two steering planes, carried on trunnions at the sides, and connected by wire strands with a drum on the deck. By turning this drum the steering planes can be simultaneously tilted to direct the machine upwards or downwards or to keep it on an even keel.

The machinery for developing and applying power is one of the most ingenious bits of steam engineering to be seen. It consists of a novel water-tube boiler, built of asbestos cloth at the sides, and the thinnest sheet steel on top. The water is contained in about 2,000 bent copper tubes, $\frac{3}{8}$ inch in diameter, heated by over 7,000 gas-jets arranged in rows. The fuel is naphtha or gasoline, which is stored in a liquid form and pumped into a vaporiser which transforms it into gas, and supplies it at a high pressure. The engines themselves are compound two-cylinder engines. They furnish 150 horsepower each which, considering that their total weight is only 600 pounds, gives the extraordinary efficiency of 2 pounds weight per horsepower! An air condenser seems to be a necessity, because the supply of water would prove a serious load. Even to drive 100 horsepower would require some 2,500 pounds of water per hour, which would be a considerable addition to a lengthy trip,

especially if undertaken for warlike purposes in a hostile country.

A reporter of the *Pall Mall Budget* some time ago visited Mr Hiram Maxim's establishment, near London, and described the trial he saw as follows:

'There was a hissing and a spluttering as some pumps got to work, and then, presently, the port propeller began to revolve with a rapidly increasing whirr-r, and the cry went up to "look out!" In a few seconds whirr-r-r-r went the starboard propeller also. The platform on which we stood rocked and

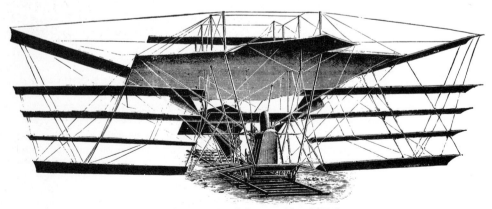

The Maxim flying machine [1894]

quivered with the vibration. A hurricane seemed to spring up, laying the hay flat far and wide, and scattering like a whirlwind the shavings in the workshop 20 yards away. Every one grabbed his hat with one hand, and clung for dear life with the other to a rail.

'Suddenly, when the tornado had reached its height, and the whole machine was shaking and straining at its anchor like a greyhound in the leash, a shrill whistle gave the order to "let go," and the huge structure bounded forward across the meadows with a smooth sailing motion, at a rate increasing up to 40 miles an hour.

'As the end of the track came in view a look of horror set in. There was nothing apparently but a quick-set hedge to arrest our wild career. A rope was stretched across the path. We crash through it. Then another, and finally we come to rest in the easiest, most graceful manner imaginable, within a few feet of what looked like perdition. Then we all laughed. It was a most delicious sensation, wiping out forever such tender memories as switchbacks, toboggans, and the seductive water-chute. It was unique, in fact, and unlike anything that the world has ever seen; for the occurrence just described represents the crude residual impressions of a first trip over the rails on Mr Hiram Maxim's giant flying-machine.

'There is a current belief that, deterred by the late accident, Mr Maxim intends abandoning further experiment. All interested will be glad to learn that this is not so. I found the men busy repairing the breakages and Mr Maxim occupied in devising improvements and means of overcoming past difficulties.'

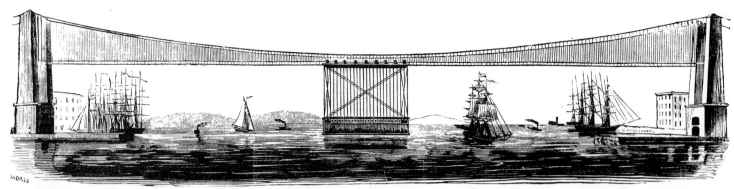

AN UNUSUAL SUSPENSION BRIDGE IN NEW YORK CITY

While plans have now been divulged to connect the island of Manhattan in New York with Brooklyn by means of a giant suspension bridge over the East River, Mr J. W. Morse has devised a bridge which permits of a much lighter construction than a normal suspension bridge and is, consequently, much cheaper to build. Mr Morse's project provides for transportation across the river in a giant platform, suspended by means of cables from a trolley running upon a gantry across the river. Measuring 40×100 feet, the platform, or traveller as it is sometimes called, has two storeys: the top floor is for pedestrians while the bottom deck is intended for horses and carriages. The car can accommodate no fewer than 5,000 passengers at each trip and it hangs at the level of the access roads, but the supporting gantry is at a sufficiently high level above the river (136 feet) to give clear passage for shipping. The traveller takes only two minutes to cross the stream, and if necessary the crossing can be made in one minute. In the course of twelve hours, 75,000 people as well as nearly 6,000 wagons and horses can be carried across.

While a normal suspension bridge requires extensive abutments and ramps to enable the road traffic to reach the bridge-deck level of almost 120 feet, Mr Morse's transporter bridge obviates the need for such provisions. The fact that the traveller hangs only 3 feet

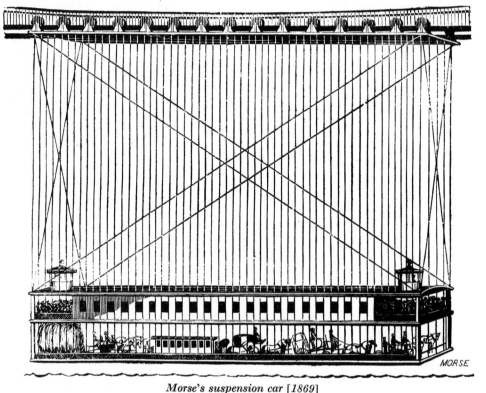

Morse's suspension car [1869]

above the water—and hence is almost at street-level—makes it easy for heavily loaded wagons to cross the river, and will also be appreciated by the workman returning home on foot after a hard day's toil in the factory or warehouse.

Machines and methods of boring the Mont Cenis tunnel [1868]

MONT CENIS TUNNEL

Except for the Suez Canal, the Mont Cenis Tunnel is the largest project being carried out by our civil engineers today. Once this tunnel is completed, the railways of France can be linked up with those of Italy, resulting in an uninterrupted connection between Calais and Brindisi, a distance of nearly 1,400 miles. The Mont Cenis tunnel and the Suez Canal will more than halve the time required for travelling from England to the Far East.

The drilling-machine shewn here was used in constructing the 8-mile-long tunnel. It is driven by compressed air; a system of pumps powered by water-turbines feeds the air to the site through metal and rubber tubes. There is an abundance of kinetic energy in the shape of fast-flowing water which rushes down from a high level. The drills are mounted on trucks moving on rails.

SUSPENSION BRIDGE BETWEEN NEW YORK AND BROOKLYN

New York, the most highly populated city of the world after London and Paris, now has more than 2 million inhabitants. It is, in fact, a cluster of four cities: New York, built upon the island of Manhattan and washed by the two mouths of the Hudson River, Brooklyn, Jersey City and Hoboken. The mouth of the Hudson which separates New York and Brooklyn is called East River; it is an estuary which is constantly navigated by innumerable ships. Until recently, steam-ships and steam-ferries were the only means of communication between the two cities, and they constituted a great impediment to the countless vessels which plied the river. During the winter, communications were frequently interrupted due to large amounts of ice floating on the water. All these inconveniences would be eliminated if the two cities were linked by a bridge. However, this had to be a bridge which would not in any way interfere with the course of the large ships passing through the river at the rate of a hundred per hour; there had to be sufficient clearance to permit them to pass below it with all sails set.

This idea eventually took shape in 1867. The contract, involving a capital of $5,000,000, was awarded to John A. Roebling, an engineer already famous for his suspension bridge across the Niagara Falls. Consisting of a single span of 1,595 feet suspended 130 feet above the water from two piers 286 feet in height, it was to be the largest suspension bridge in the world.

Unfortunately, John A. Roebling did not live to see the foundation-stone of this vast project being laid for in 1869 he was already dead due to an accident which befell him during the surveying work preceding the actual construction. A worthy successor was found in his son Washington A. Roebling but he too failed to live to see the work completed: in 1871, he was afflicted by an illness brought on by his continuous presence in the chambers filled with compressed air used for working under water. Although partly paralysed and suffering intense pain, Roebling continued to direct the work from his sick-room, faithfully assisted by his wife who had acquired the necessary engineering knowledge for the purpose. Following her husband's death, this remarkable woman succeeded him in the direction of the work which, under her guidance, was successfully completed without any major accident.

The supporting piers have been sunk into the bed of the river with the aid of caissons, one for each pier. Each caisson—these are, in fact, giant diving-bells—was built up on shore from heavy timbers connected by iron bolts and rendered air-tight with pitch and tar. The cast-iron bottom rim was formed into a wedge-shaped cutting-foot and pressed into the river-bed by the weight of the caisson. Each of these structures is 75 feet in length, 105 feet in width and has a working-chamber 9 feet high. The timber walls are as much as 8 feet thick because the caisson must be capable of bearing a weight of more than 80,000 tons.

Upon their completion, the caissons were floated out to the site planned for the piers. Here, heavy weights were placed upon them so that they were almost completely submerged, thus forming artificial islands on which the granite blocks for the piers could be securely built. The bottom chamber was kept full of air by six pressure pumps driven by steam.

Pressure chambers serving as air-locks were used to permit the workmen to enter or leave the working-chamber without changing the air pressure of between 3 and 4 atmospheres. The navvies having entered the air-lock, its door was closed hermetically and the air pressure was then gradually raised by opening a cock which was connected to the working-chamber. All sorts of provisions had to be made for the more than 200 men working in the dark space below the water-level.

An overall view of the Brunton machine intended for driving the tunnel between England and France [1874]

Mr Farrington, chief supervisor of the construction of Brooklyn Bridge, was the first to cross the span between the two pylons, on 25th August 1876

Water for drinking and washing was piped to the working-chamber by a special conduit. The whole chamber was lit by fifty-six gas flames. Since ordinary lighting-gas gives a poor flame in compressed air, oxyhydrogen gas was used as a source of light.

The caisson on the Brooklyn side was brought into position on 2nd May 1870 and filled with concrete on 11th March 1871, i.e. thirteen months later, in spite of a fire which could be extinguished only by flooding the whole caisson with water. An increased fire hazard was found to exist in compressed air, and for that reason the interior of the caisson on the New York side was entirely clad with iron. This caisson pushed through the upper layers of soft silt and came to rest on the river-bed 100 feet below water-level, the first depth at which solid rock was encountered. Together with the tower built upon that foundation, the total height was 360 feet. Including 1,200,000 cubic feet of masonry-work, the total weight comes to nearly 100,000 tons!

To provide anchorage for the four main cables designed to carry the bridge decks, large masses of granite were placed in excavations on the river-banks 860 feet away from the centres of the towers, each granite block weighing 60,000 tons. Each of the steel wire cables consists of 5,282 single wires $\frac{1}{8}$ inch thick laid parallel to each other. They are divided into 19 strands of 278 wires, 3 inches in diameter. Together, they form the large cable which is 20 inches thick. The strands are united by means of steel wire wound helically round them. Over 3,000 feet in length and with a circumference of 5 feet, these cables were too large to be transported and had therefore to be manufactured on the site itself, wire by wire, strand by strand. The main problem was the stretching of the first cable between the piers. First, it was placed on the bottom of the river, ready to be hauled up when there was no shipping traffic on that part of the river.

On 14th August 1876, a gun-shot marked this auspicious moment and this cable, hoisted aloft by means of steam-engines, formed the first link between the two cities. On 25th August Mr Farrington, Chief Supervisor of the project, completed the first journey through space, hanging from the two tautly drawn wires which were invisible to the spectators observing the event from a distance; they were at a loss to understand by what means this man, a new type of air traveller, propelled himself through the air, rocked to and fro by the wind. The original nature of this audacious trip amazed the numerous spectators and encouraged all those who were actively engaged in the great enterprise. The excited population of the two

cities made this journey into a veritable triumphal tour.

Now the men, who had at first lived and worked like fish at the bottom of the water, were to live and work like birds high in the sky. In platforms and cages they glided upwards along the steel cables, 200 to 300 feet above the waves, stretching and securing the more than 10,000 steel wires which had to be joined together as each one was only 1,000 feet in length.

The main difficulty in this giant undertaking lay in giving the proper direction and curvature to the suspension cables to ensure that each of them would carry an equal part of the bridge's weight. This had to be mathematically calculated beforehand, with due regard to the violent storms which frequently rage in this area. It took three weeks of waiting for an entirely calm and windless day until the correct curvature of the first cable could be established. Moreover, since atmospheric influences such as sun, shadow, cold, heat and wind greatly affect the span, the work of adjusting the cables could be done only before sunrise or after sunset or on cloudy and foggy days.

The work was frequently interrupted by ice forming on the cables during the cold season.

The weight of the bridge structure suspended from the cables is 14,830,000 pounds and the maximum load under heavy traffic 3,036,000 pounds making a total of 17,866,000 pounds. Each individual cable itself weighs 1,905,000 pounds.

The steel floor of the bridge is 80 feet in width and accommodates a double railway track, two roads for coach traffic, and a sidewalk for pedestrians. The total cost now exceeds $15,000,000. Engineers are now engaged in constructing a cable railway on the bridge to move railway carriages and trucks across it; no locomotives will cross the bridge. This requires a $1\frac{1}{2}$-inch-thick cable running on 500 discs and driven by two steam-engines over drums 13 feet in diameter.

The honour of first crossing the bridge upon its completion was granted to Mrs Roebling, the Chief Engineer's widow who so successfully followed her husband in directing and completing an enterprise which the Americans are fully justified in calling one of the wonders of the nineteenth century. Thus, the honour bestowed upon Mrs Roebling is a

Brooklyn Bridge under construction: braiding the suspension cables [1876]

legitimate homage to the merit and talent of this courageous woman, and at the same time a protest against the old superstition that any bridge which is first crossed by a woman will be under a curse.

THE GREAT RAILWAY BRIDGE OVER THE FIRTH OF FORTH

The construction of the great railway bridge to cross the Firth of Forth is one of the grandest works of modern engineering. It was designed, for the North British Railway Company, by Sir John Fowler and Mr Benjamin Baker, has been five years in actual progress, and will be completed in the autumn of this year. The width of the estuary in this part is reduced by the peninsula of North Queensferry to a mile and a half, and on the south shore the water shoals rapidly, with a bed of boulder clay and a very deep stratum of mud, but the Fife shore is an almost perpendicular cliff, and the intervening islet is a rock in the centre of the deep channel, with 200 feet depth of water on each side, and with a strong tide current sweeping up and down on each side. It was impossible to erect piers anywhere but on this islet; hence the bridge must rest on three main piers

Progress of the Forth Bridge [1889]

Construction of the south pier of Forth Bridge, 15th April 1887

which serve to relieve the balance arms of the cantilever girders, and to connect the bridge with a long approach viaduct.

A cantilever is a girder supported at one point, its overhanging extended part being balanced by its weight at the other end. This engineering device is the most novel feature of the Forth Bridge. The main spans of the bridge are to be upheld over the deep-water channels by the projecting ends of cantilever girders, with connecting central girders over about one-sixth of the span. Each cantilever girder is a complex structure framed of four vertical columns, standing not parallel, but from a wide base narrowing to the top.

STRAUB'S PLAN FOR A SUB-RIVER TUNNEL

The illustration represents a novel form of sub-river tunnel, more especially designed for the Hudson and East Rivers at New York City. It is proposed to construct such tunnel of several long sections of steel tubes, about 18 feet in diameter, with heavy strengthening flanges passing round the tubes at intervals of 5 or 6 feet, while there are also ribs running lengthwise of the tube. These tube sections are to be constructed above ground and lowered into a prepared line of way previously dredged or otherwise made in the river-bottom. Figure 1 shows a cross-section of such tunnel in position, with its top weighted by stone and cement covering, to hold it firmly in position. [1889]

A HELMET FOR PROTECTING THE HEAD

Now that there is such an increase in travelling by sea and land we find it surprising that inventors pay so little attention to rescue equipment. An American, Mr Francis P. Cumberford, has invented a helmet which can be of great service in case of shipwreck.

Hydraulic shield used in the St Clair River Tunnel [1896]

Proper position for floating [1881]

Cumberford's head protector [1878]

Novel swimming device [1880]

A LIFE-PRESERVER

Mr Traugott Beek of Newark, N.J., in the United States of America has invented a life-preserver, the top part of which consists of a floating-buoy in which the wearer has freedom to move his head and arms about. It provides those unfortunate enough to be shipwrecked not only with sufficient power to float but also affords them complete shelter. A month's supply of food and drinking-water can be stored in the upper section. The cover can be closed when high seas are running, adequate visibility then being provided by a window, while the occupant can breathe through a curved pipe. The preserver is fashioned of waterproof sailcloth secured to circular metal tubes, while the watertight trousers and gumboots with metal bands provide protection against injury from rocks and voracious fish.

The helmet is made of rubber and fits tightly round the whole head and the neck. It is fitted with two glass windows for the eyes and at the rear there is a hollow space which is perforated and shaped in such a way that air can be sucked through it to permit respiration, while any water which penetrates into the cavity will drain away without entering the remainder of the helmet. There is a hole in the headgear at the place where the mouth is situated and this can be closed by means of a rubber stopper, which may be replaced by a bell-shaped speaking-trumpet to amplify the speaker's voice. The helmet may also be employed in ships to afford protection against rain or snow. A lighter type can be worn by ladies as a bathing-cap.

A SWIMMING DEVICE

Monsieur A. Gamonet of Lyons, France, has invented a practical apparatus for persons unable to swim. A handle operated by the bather's right arm actuates a rearward-acting propelling device which is constructed so as to collapse when pulled forward, and to spread open when pushed back, resulting in a forward movement. Inflated India-rubber bags afford the necessary buoyancy to the swimmer.

Beek's life preserver [1877]

THE NEW BESSEMER SALOON STEAMER

As our engraving readily conveys, the Bessemer saloon ship has as its distinctive feature a saloon that maintains an even level, whatever the motion of the vessel. This revolutionary principle is a joint invention of Mr H. Bessemer, author of notable improvements in the iron and steel manufacture and Mr R. J. Reed, C.B., late Chief Constructor of the Navy.

The Bessemer saloon forms by far the finest cabin that has ever been fitted in a ship. Its great size and height enable it to be completely ventilated, unlike the ordinary cabin between decks which is so unpleasant that ladies and delicate persons endure the worst weather on deck rather than accept shelter in it. But the greatest advantage is that the saloon, being virtually isolated from the hull of the ship and subject to the action of Mr Bessemer's hydraulic levelling apparatus, is designed to remain absolutely unaffected, no matter how much the vessel may roll. Mr Bessemer's approved statement indicates that sea-sickness among passengers is completely eliminated, all sense of pitching and rolling being so small as to be inappreciable. Mr Bessemer's hydraulic apparatus is an established certainty, and not a matter of speculation, and it will always insure the floor being kept level.

Our engraving shows the equanimity of the passengers in the saloon area, whereas those on the ordinary deck are subject to the usual discomforts. The illustration was prepared, however, before the ship's first voyage in rough weather, when the device proved not entirely satisfactory. [1874]

PREVENTION OF SEASICKNESS

A life on the ocean wave is a fine thing in poetry, but in practice, to those whose stomachs are sensitive to the motion of vessels, it is often a very sorry experience. Many and various remedies, and as many prophylactics as remedies, have been proposed, among which the most efficacious is to stay at home, but the latter, unfortunately cannot always be done. The inventor of the device illustrated has, however, undertaken the task of providing a remedy for sufferings of seasickness. If successful in operation the discomforts of a sea voyage to many will be overcome.

The invention provides the staterooms, cabins, saloons, etc., of vessels with couches, sofas and the like, suspended in such a way as always to maintain a horizontal position, no matter how much the vessel may pitch or roll. The couches are preferably made in a

The floating steam fire engines of New York: burning of the ferryboat Garden City [*1884*]

circular form, and suspended on oscillating hangers, the hangers being adjusted on the principle in which the mariner's compass is suspended to keep it constantly level. The hanging couch may contain a centre table, and other small articles of furniture.

A NEW SURF BOAT

The engraving represents a novel surf boat recently patented by Mr Richard H. Tucker of Wiscasset. The boat is circular in form, with convex upper and lower surface, and its entire interior forms a reservoir for holding compressed air to be used in the propulsion of the boat. The propelling device is very simple. It consists in air-nozzles projecting towards the stern, one being placed in each between the keels, of which there are several. The air-nozzles are provided with valves which are operated from the deck. The boat is steered by closing the air-valves on one side or the other as may be required.

The boat is not designed for long distances, but it is claimed that it has propelling power sufficient for ordinary requirements. It certainly contains no machinery which can become impaired either by use or rest, and it possesses sufficient buoyancy and is of the proper form to maintain its position in the water.

Newell's oscillating sofa, table and couch for vessels [1870]

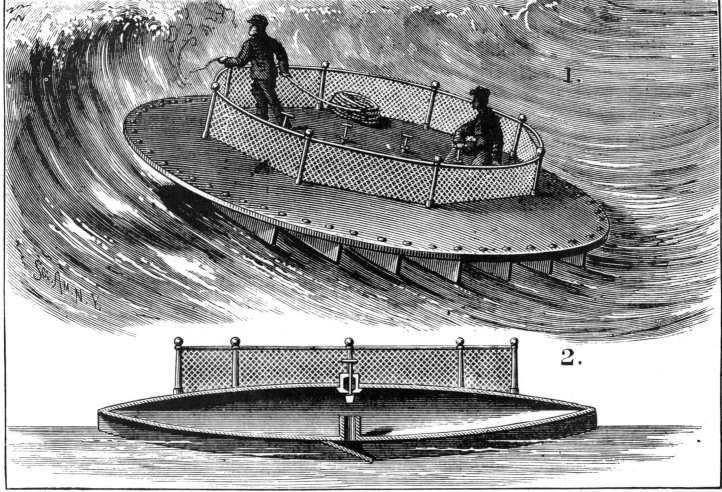

Tucker's surf boat [1879]

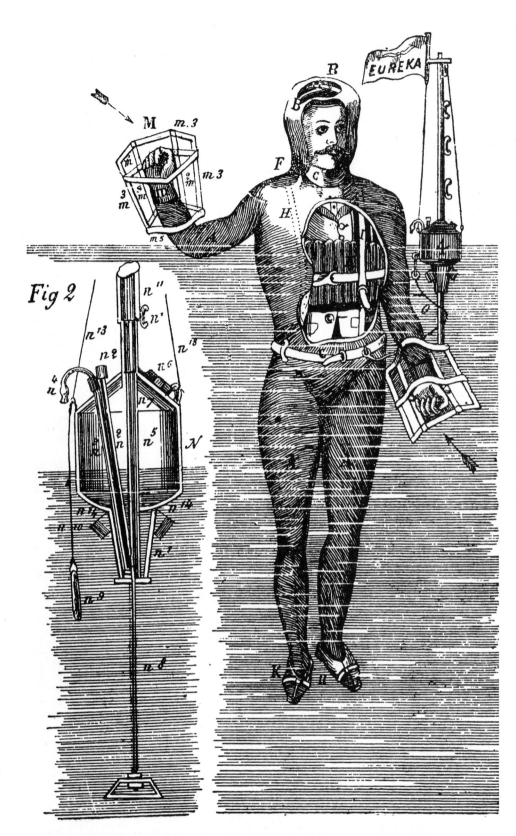

Fig 2

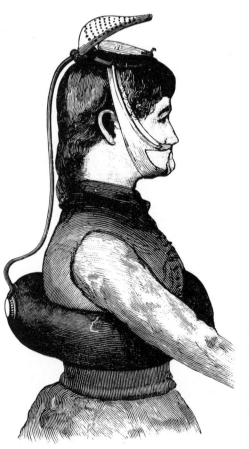

An improved life preserver [1886]

Round the hands, a rubber-clad framework may be seen which assists the user in propelling himself through the water. The demonstration was a complete success and Captain Stoner, the inventor of the apparatus, received hearty cheers from the party aboard the steamer. [1869]

GOUBET'S SUBMARINE VESSEL ORDERED BY THE RUSSIAN NAVY

In 1881, following a series of experiments lasting for more than twelve months, the Russian Navy ordered a few submarine boats as designed by Monsieur Goubet, the Parisian engineer. Propelled by an electric motor, the craft holds ten accumulators which also feed a powerful electric searchlight mounted in the bows. By permitting sea-water to flow into the lower compartment, the boat will submerge; after the water has been pumped out by hand, the submarine vessel will come to the surface again. It can be propelled by hand in the event of a breakdown occurring in the electrical installation. In case of emergency, or should the vessel sink, the leaden keel attached to the underside of the boat may be jettisoned, thus causing the craft to come to the surface under its own buoyancy. The dished top plate on deck is intended for bringing a floating mine with a powerful explosive charge under an enemy vessel.

A NOVEL LIFE-SAVING DEVICE

During a boat trip on the Hudson River near New York, a demonstration has recently been given of a new life-saving apparatus which will enable a shipwrecked person to float for many days in the water in an erect position, thoroughly protected from cold and wet by a helmet and rubber suit under which a buoyancy-belt is worn. A floating magazine shewn in Fig. 2 forms part of the equipment, its purpose being to provide the person afloat with drinking-water, food, reading-material so that he may read the news to pass the time, cigars, a pipe and tobacco, as well as torches and rockets to make known his position. The small flag-pole with the 'Eureka' flag facilitates the location of the shipwrecked person.

Goubet's submarine vessel for the Russian Navy [1886]

Gas engine explosion on 16th December 1886 at Asnières, France, in which Just Buisson and his young assistant lost their lives.

A SHIP ON WHEELS

In the shipyards of Saint-Denis in France, a paddle-steamer is now being built which is entirely different from any existing type of ship. Designed by Monsieur Bazin, the craft consists of a platform with a sharp edge in front supported by large, hollow wheels which keep the platform about 20 feet above the water-level. To transmit the rotating power to the paddle-wheels, steel shafts 28 inches in diameter supported in sturdy bearing-blocks penetrate through and below the platform.

The engine-rooms, stokeholds, passengers' cabins and the various service-quarters have been built upon the platform. Propelled by two screws, the ship will roll over the water with a minimum of friction.

Monsieur Bazin has perfected his construction by fitting a continuously operating, hydraulic rudder consisting of an upright column at the ship's stern, which is controlled by the steersman. From this column, a powerful water-jet is forced into the water and this will govern the ship's movements by reacting on the sea. The propulsive power of this rudder will—either fully or in part—be used in navigation, and even with the engines at a standstill it will serve to steer the ship to her berth at a speed of a half or quarter knot.

———

CHAPMAN'S ROLLER VESSEL

On 20th November 1895 the *Call*, a San Francisco newspaper, published an article under the title 'Vessel to roll on the water'—giving an account of a ship which, it was claimed, could compete with the fastest of trains for speed. The inventor, Mr Chapman, gave the following account of his vessel: the hold, bridge and passenger cabins seem to be squeezed between two gigantic rollers journalled [i.e. on bearings] in gangways on either side of the ship. The interior of each roller is equipped with a narrow-gauge track on which a locomotive driven by electricity can run.

As soon as the locomotive is set in motion, the huge drums start rolling, moving the ship in a forward direction. Very high speeds may be attained. Mr Chapman even claims that the top speed of his vessel will not be much less than that of a modern, fast train so that the crossing of the Atlantic between New York and Britain may take only three days or even forty-eight hours while the passengers will also be virtually free from sea-sickness.

Chapman's roller vessel [1897]

Bazin's ship on wheels [1890]

A suggestion in canal boat propulsion [1889]

A PLAN FOR A 1000-FOOT-HIGH TOWER

One of the most original and daring enterprises of our times is that of the French engineer, Monsieur Eiffel, who intends to erect a huge tower, 1,000 feet high, which will permit the visitors to the forthcoming World Exhibition to have an overall view.

The highest building in the world is Cologne Cathedral which is 523 feet high. For buildings beyond that height, it is necessary to have recourse to steel. This metal is the only material capable of withstanding not only the enormous vertical pressure of such a colossus, but also the tremendous impact of the wind force during a storm.

In the tower planned by Gustave Eiffel, the main structure consists of four styles which form the edges of a quadrilateral pyramid with slightly concave sides. The curvature has been calculated mathematically and by virtue of this alone the tower will be capable of offering adequate resistance to high winds. This is the main characteristic of the grandiose plan.

A glass dome on top of the tower will allow the visitor to admire the immense panorama which unfolds at his feet. To reach the top, he will avail himself of one of the four *ascenseurs* to be accommodated inside the columns and provided with various devices to ensure the utmost safety to the passengers. A meteorological and astronomical observatory will be built at the top of the tower, permitting, for example, spectroscopic observations to be made with a far greater accuracy than at ground-level, due to the greater purity of the air at that level. Eiffel finally proposes to place powerful electric light sources on top of the tower and thus provide bright illumination over an extensive area. Full of admiration for the enterprising spirit and vivid imagination of the French, we look forward to the materialisation of this audacious plan.

THE EIFFEL TOWER COMPLETED

The 1,000-foot-high tower, designed by Gustave Eiffel, is completed! And that in spite of the storm of indignation which it raised, including the petition signed by many well-known Frenchmen who contended that a nation which had at all times upheld the banner of civilisation and refinement, was giving itself a slap in the face by erecting a monument which pursued no other aim than that of publicity, a structure to dwarf all the splendid buildings of which Paris was so rightly proud, and as such was nothing less than a tremendous profanation of the arts.

Be that as it may, this colossal structure designed by Gustave Eiffel, who has already won great fame for his bold bridge constructions, now rises proudly from the Champ-de-Mars in Paris. Consistently refraining from using scaffolding, he adopted an entirely new system: working from the piles, the arches were built up piecemeal until the two parts were joined in the centre. The design, calculation and drawing of the tower project alone required a staff of forty persons; over 3,000 detail drawings were made.

The tower was to be erected upon a layer of clay-soil enclosing a seam of sand which tapered off towards the River Seine; the subsoil was composed of chalk. The foundation employed was of the pneumatic type: steel caissons were filled with a bed of concrete 50 feet long, 22 feet wide and 7 feet in depth upon which masonry was placed until it rose just above ground-level. Initially, the bottom of this foundation was slightly higher than what was later to become the base: the space between was excavated by workmen so that the caisson sank into the soil under its own weight. The spoil was removed through vertical shafts, while the navvies worked under electric light and in compressed air. The tower is built upon sixteen such piles, made of concrete and masonry and covered with coping-stones able to withstand a pressure of more than

Lifting one of the main columns by hydraulic jack [1889]

Sectional view of the Otis elevator [1889]

form there are several rooms intended for those wishing to engage in scientific research; these rooms are closed to the public.

Various lifting devices carry the visitor upwards. The Combaluzier lift goes to the first platform. Endless chains, composed of a series of hinged bars, move on guide-pulleys driven by two powerful hydraulic engines. These chains are fastened to the sides of a compartment which can convey a hundred persons to the first platform within a minute. The Otis elevators leading to the second platform are capable of holding fifty passengers but travel at double the speed: 7 feet per second. In a huge cylinder, 35 feet long and more than 3 feet in diameter, a piston is moved under a pressure of 170 pounds per square inch by water which is stored in reservoirs on the second platform. The

The Edoux hydraulic lift [1888]

2,500,000 pounds per square foot.

Up to a height of 100 feet, the structural work presented little problem since a crane of that height was used to lift the steel members into position. The sections were provisionally joined together with screwbolts which were later to be replaced and secured with rivets. At high levels, four creeper cranes each weighing 26,500 pounds were used. At a height of 200 feet the four columns, each forming one edge of the quadrilateral pyramid, were joined together by 22-foot-long, 75-ton girders to form the base for the tower.

From this first platform, the work proceeded in the same way as before. A railway track was laid on it and the steel members were hoisted up by steam-driven cranes. At a height of 350 feet, the second platform joined the four main styles together. There, a slight discrepancy in height was found: two of the columns were one-quarter of an inch longer than the other two. Proof indeed of the

scrupulous accuracy with which the work had been carried out!

Beyond the second platform, however, no base could be found on which to mount four cranes. Eiffel therefore decided to affix two cranes to vertical frames, so that one of them would serve to counterbalance the other. The crane could travel 30 feet upwards along this frame; once this level had been reached, a second frame was fastened on top of the first one, and so on. It took only forty-eight hours to make this change which involved a weight displacement of nearly 50 tons! On 30th March 1889, two years after the commencement of the structural work, the entire tower was completed. It was inaugurated the next day amid much festivity.

The top section of the tower, in elevation a square with an edge of 50 feet, contains a gallery accommodating 800 persons which can be shut off with glass walls, a necessary precaution in view of the strong winds prevailing at that height. Above this plat-

compartment, suspended from the loose end of an enormous tackle, is lowered 12 feet for every foot the piston travels. The Edoux elevator carrying the visitors to the top in two stages is a vertical lift moved by the pressure of 700 cubic feet of water stored in a tank at a height of 1,000 feet. The water required for the various elevators is carried upwards by steam-operated pumps producing a power equal to 300 horsepower.

There are people who doubt whether the cost of the Eiffel Tower will ever be recouped. In this our nineteenth century, people are ever ready to question the practical value or the financial advantages to be gained from civil engineering projects, and they find no satisfaction in the enthusiastic comment of an engineer: 'But, Sir, it is a masterpiece without an equal!'

Well, capitalist, let the Financial Committee of the Exhibition who granted a subsidy of £450,000 bemoan their lot. We for our part pay tribute to the genius of the man who succeeded so brilliantly in overcoming all the technical problems which confronted him.

He who can build an Eiffel tower, useless monument though it may be, can do still more; let us therefore continue to pay honour to such men as these!

HYDRAULIC LIFT

Leo Edoux, a Frenchman, has designed and constructed a passenger lift, which is pushed upward by a plunger in a cylinder, the necessary power being provided by water under pressure.

Amiot's stair climber [1889]

AMIOT'S STAIR-CLIMBER

A device has been designed by a Frenchman, Amiot, to do away with the fatigue of climbing stairs. It can easily be installed in dwellings and offices where a lift would be too cumbersome or too costly. This invention must indeed be a boon to those who are feeble on their legs. It consists of two metal rails, one placed above the other, and a platform with wheels capable of moving up and down this track under motive power obtained from an electric motor or other suitable machine to which it is connected by means of a chain or cable. Since piped water under high pressure is available all over Paris, a hydraulic motor of the type already used for many other purposes, may be employed to advantage here.

Rolling staircase at the Paris World Exhibition [1900]

A MACHINE FOR SENSATIONAL EMOTIONS

Monsieur Carron, an engineer from Grenoble in France, has devised a machine which will delight the lovers of sensational emotions. In planning this machine the inventor had in mind those persons who enjoy the unnerving sensations experienced, for example, in high swings or extremely fast sledges as they hurtle headlong over mountain-slopes. In order to evoke even stronger emotions than these he intends to allow the public to participate in a free fall of 325 yards. The possibility for this is provided by the Eiffel Tower which is of the height just mentioned. If Monsieur Carron's calculations are correct, the speed attained at the end of a free flight such as this is 84 yards per second, corresponding to about 172 miles per hour, a speed at which no human being has ever travelled as yet. A comparison may be provided by the fact that our fastest express trains cover a distance of about 32 yards per second, or approximately 65 miles per hour.

Making a free fall such as this will indeed be a vertiginous experience. It is easy to fall

325 yards, but it has hitherto been doubtful whether one could do this and survive. This problem has been solved by the inventor. He has designed a cage in the shape of a mortar-shell containing a round chamber some 13 feet high and 10 feet in diameter in which fifteen persons can sit extremely comfortably in well-upholstered armchairs arranged in a circle. The floor is formed by a mattress with spiral springs 20 inches high. The bottom half consists of concentric metal cones which provide a further measure of resilience. The total height of the apparatus is almost 33 feet and its weight, inclusive of the electric lighting, 10 tons.

It is intended to drop this gigantic shell from the top of the 325-yard-high Eiffel Tower. It will be prevented from being dashed to smithereens by falling into a water-filled pond shaped like a champagne-glass. This pond will be 60 yards deep with a maximum diameter of 54 yards. The water will serve as a shock-absorber. Mr Carron assures us that by virtue of this, and because of the springs inside, the shock felt by the occupants on landing will be in no way unpleasant. When they have got out, the giant shell can again be hoisted to the top of the Eiffel Tower to permit another group of adventurers to experience the thrills of a free fall. According to the inventor, the shell can be operated profitably at a fee of twenty francs per passenger per trip which is by no means an excessive charge for such a vertiginous experience as this promises to be. [1891]

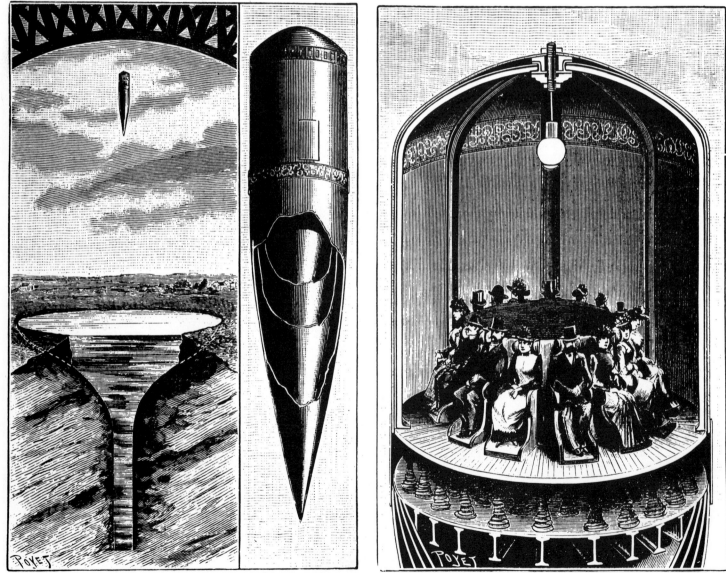

Left, the water-filled pond: centre, details of the shell construction: right, interior of the cabin accommodating 15 passengers [1891]

Midget electric train plying between M Gaston Menier's kitchen and dinner-table [1887]

The electric train serving the fish and the sauce [1887]

ELECTRICITY

EDISON'S INCANDESCENT LAMP

In 1800 the British physicist Sir Humphry Davy received from the Royal Institution a battery of 2,000 volta elements in token of appreciation for his research work. During his tests with the battery, Davy discovered that a brilliant, blinding light could be produced at the tips of two charcoal rods by passing an electric current through them. But not until the second half of our nineteenth century—when dynamos became available for generating electricity—was the arc-light, as it was called, used for lighting purposes, among other things, at Trafalgar Square in London and during the siege of Paris in 1870.

However, the arc-light was far too brilliant for the lighting of dwellings and offices and efforts were therefore made in many countries

MACHINE SHOP.

Edison's electric generator [1880]

to subdivide the electric light, i.e. to divide the light among a large number of small lighting units. Since all these attempts remained fruitless, however, the work was finally abandoned. 'The sub-division of the electric light is a problem that cannot be solved by the human brain', was the verdict given in 1878 by a committee of prominent scientists appointed by the British Parliament. This view was endorsed by their American colleagues.

Thomas Alva Edison, the American inventor, regarded this as a challenge and decided that he would solve the problem. He began studying all the available literature about electric lighting and made a series of tests with a thin platinum wire which he melted into a glass globe from which he evacuated all air by suction. Edison succeeded in producing a faint light with the platinum wire which, however, melted away when he increased the voltage. This caused Edison to turn away from this precious material, and he undertook experiments with filaments of carbonised matter. In a nickel mould placed in an enamelling-stove, the inventor carbonised a large variety of materials: baling-twine, coconut fibre, a piece of angling-line, wood, a violin-string, crocodile leather, and many more besides. But all these attempts failed.

Early in October 1879, Edison decided to resort to ordinary cotton sewing-thread; prepared with coal-tar and shaped like a hair-pin, it was placed in its mould in the enamelling-stove and heated for five hours. During the attempts to enclose it in the glass bulb, however, the brittle thread broke. A second filament was prepared; it also broke. A third, a fourth and many more threads were carbonised, but however carefully they were inserted into the bulbs, they were all too brittle and broke into pieces.

But on the 21st October the patience of Edison and his collaborators was finally

Edison's electric lamp [1880]

An unusual test with electric light [1884] (see page 91)

rewarded: they succeeded in melting a carbonised thread into the glass globe which, having been emptied of all atmospheric air by means of a mercury suction-pump, was connected to the electricity supply from a dynamo. A clear, steady light was emitted by the carbon thread, and this burned unwaveringly for a quarter of an hour, half an hour, an hour. . . . Small wonder that this memorable event gave rise to an enthusiastic outburst of unbridled joy, after thirteen months of seemingly fruitless toil and endless disappointment. All the people in the neighbour-

hood were invited to come and see the phenomenon, and everyone watched in awe as the fragile filament spread its lustre while resisting an amount of heat that would have melted threads of platinum and even of iridium.

'It burned like an evening star for forty-five hours,' to quote Edison's own words, 'then the light went out with an appalling, unexpected suddenness.'

The news that the great experiment had eventually been crowned with success, sped along the telegraph-wires of the world to

Edison's new dynamo [1882]

announce to the four corners of the globe that a new light had been born, a light that was gentle and mild yet bright as day itself. A light without noise and odour, a light without flame and hence free from the hazard of fire. No matches were needed to light it; it produced only little heat, did not vitiate the air and did not flicker. It would be cheaper than the cheapest oil, cheaper even than lighting-gas. . . . This was to become the ideal source of light, and its benefits were for all to share. However, despite the newspaper reports that Edison, the wizard of Menlo Park, had invented an electric lamp that could burn for dozens of hours at a stretch, doubts as to their veracity continued to be raised, and there were newspapers and persons who described Edison as a swindler and an impostor. The inventor therefore decided to stage a public demonstration.

On New Year's Eve, hundreds of visitors were conveyed to Menlo Park by special trains. The city fathers and leading scientists and industrialists had received an invitation from Edison to join him in a New Year party.

When the guests arrived, there was annoying darkness, but all of a sudden hundreds of lights shone forth . . . a sight so brilliant and fairy-like that it was termed 'a heavenly spectacle' by the Press.

With this demonstration, Edison carried the day: the gas companies' shares plummeted even more steeply than after the first reports of Edison's invention, whereas the value of Edison Electric Light Company stock rose from 106 to 3,000.

To Edison, however, this was only the beginning. He was now faced with the problem of producing this 'lamp for everyone' in large quantities. And then? On what energy source were these lamps to burn? How were the electricity supply-lines to be fitted and made safe for the public? What means were to be adopted for measuring the consumption of electricity, and how were the lamps to be fitted?

First and foremost, the filament had to be improved. Again, thousands of carbonising tests were carried out in which the best results were obtained with bamboo. Thousands of incandescent lamps were made of bamboo fibre imported from Japan.

Having carefully studied the map of New York City, Edison selected a quarter south of Wall Street to be converted into the world's first city district to be lit by electricity. His assistants reconnoitred every street and every house, noting all the data about the gas-lighting systems and the length of pipes in the houses. Edison then designed all the equipment necessary to replace the lighting-gas with electricity: switch-boxes, fittings and sockets, wall-switches, fuse-boxes and electricity consumption-meters. Virtually nothing existed that he could use for his

Carriage lit by electricity [1885]

purpose. For accommodating the bulbs, he devised a screw socket. The main problem was to design a generating station capable of supplying electricity to 13,000 lamps.

For this Edison bought a block of four

The electric torchlight procession in New York [1884] (see page 91)

storeys, three of which are shewn in our
engraving. It houses eight steam-boilers and
an equivalent number of steam-engines with
a combined power of 2,000 horsepower and
driving eight dynamos designed by Edison.

In the meantime, workers started breaking
up the streets to dig trenches for the con-
duits through which the electric cables would
run.

On 4th September 1882 the generating

station was brought into operation by Edison
himself in the presence of many authorities
and other invited guests. 'I was extremely
nervous,' said Edison later, 'when I had
started the thing up, and for some time I felt

The three lower storeys of Edison's central station for electric lighting in New York [1882]

The dynamo room of Edison's electric lighting station [1882]

uneasy for I did not know what would happen. The steam-engines and dynamos made an ear-splitting din ranging from deep groans to fearsome screams and the place seemed full of sparks and flames of every imaginable colour. It was as though the gates of Hell had suddenly opened up!'

The moment the dynamos began to rotate the current flew through the wires, flowed along below the streets and penetrated into the houses, shops and offices. At first they looked like red-hot nails but when all the dynamos were rotating at full speed many thousands of incandescent lamps shone with a calm, clear light to the accompaniment of cheering and delighted cries on the part of the inhabitants. The incandescent lamp was no longer a hothouse plant from a laboratory —it had become a useful, everyday article that could burn in the living-rooms of the citizens of New York. And so Edison had achieved his principal goal: to create an electric lamp which would be within the reach of all. Not a lamp for the millionaire, but for the ordinary man.

'I shall enter into competition with the gasworks,' said Edison. 'The electric cable will be laid in every street and the wires which will be led into the houses will bring not only light, but motive power and warmth as well. With electricity you will be able to drive sewing machines, shoe-cleaning machines and washing machines; you will be able to light your house and cook your food.'

Even if the expectations of the great inventor about bringing power and heat into houses along this wire remain but a dream, his system permitting us to light our homes with cheap electric light will nevertheless remain one of the most beneficial discoveries for which the world must be forever in his debt.

TWO UNUSUAL TESTS WITH ELECTRIC LIGHT

The original procedures which many Americans are accustomed to adopt when they want to attract the public's attention in matters dear to them are well known. Countless peculiar advertisements in their newspapers prove that the Americans are past masters in matters of publicity.

The fact that even large companies in the United States are not loath to employ spectacular advertising is proven by a few examples in which the leading role is played by the electric light.

To attract the attention of the visitors to the Electricity Exhibition held at Philadelphia, the Edison Electric Lighting Company devised the original idea of having their address-cards distributed by a tall negro wearing a helmet to which an Edison lamp had been fastened. The lamp on the helmet was connected to two wires concealed below the negro's garments which led to two copper plates fitted under the heels of his boots. Other copper plates embedded in the ground were connected to the poles of a dynamo-machine used for the lighting

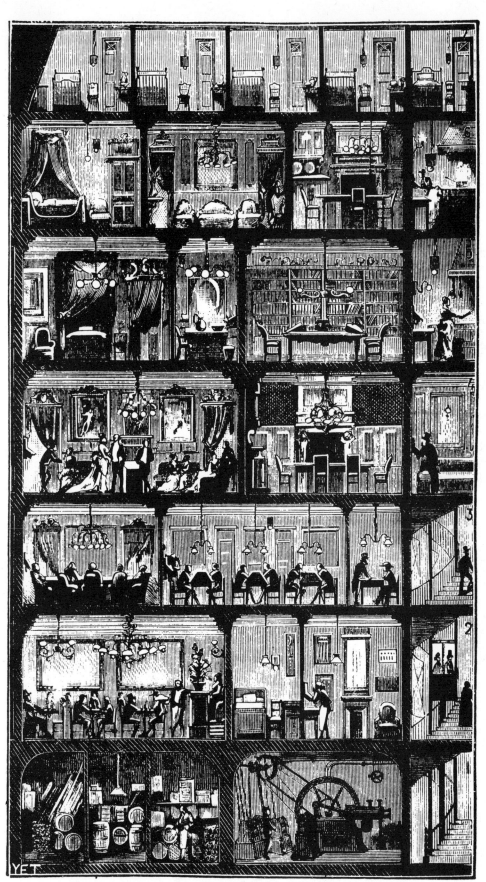

House fitted with electric lift and lighting, the generating equipment being accommodated in the cellar [1884]

system. Whenever both the negro's heels were resting on two of these plates he could, by a barely discernible movement, connect or disconnect the flow of electricity to the lamp on his head, thus lighting or extinguishing it without using his hands which were needed for distributing the address-cards.

It will be understood that various nervous

people were frightened by the sudden appearance of the brilliant electric light which attracted such large crowds that the negro had to keep walking on in order to avoid creating an obstruction. This joke was taken so seriously by some Americans that they inquired about the cost of this device which, they said, was just what they needed for an inspection tour round the house.

The second advertising scheme for electric light was used by the Edison Company in a publicity campaign for the presidential election. This extensive trial demonstrated that a completely equipped installation for large-scale lighting can be fully mobile without interrupting the current or impairing the illumination level of the lamps. It was a parade such as the world had never seen before in which the usual traditional torches and Chinese lanterns were replaced by lamps lit by means of electricity.

The front part of a large wagon was occupied by a 200-ampère dynamo-machine, which was placed in front of a steam-engine capable of delivering 40 horsepower. This power was transmitted to the dynamo by a drive-belt. To ensure the production of a sufficient quantity of steam, the machine of a large fire-brigade engine was used, the boiler of which can rapidly generate a head of steam even when the engine is in motion. Following the steam-engine and the fire-tender came two water-tanks holding a combined total of 1,000 gallons. These were connected to a feed-pump by means of tubes, and between them again were two coal-trucks. The wagon carrying the dynamo and steam-engine was drawn by six horses harnessed one behind the other and controlled only by the voice of the driver.

From the wagon there ran two 1,300-foot-long ropes enveloping two insulated copper wires. These formed the link uniting all those taking part in the parade. The ropes were tapped every 5 feet: two wires connected them to a lamp fastened to the helmet worn by each of the participants. The lamps were each of 16 candle-power. Each horse's harness also carried two lamps, and a further twenty-four lamps were used to illuminate the wagon. The leader of the parade, riding on horseback, had a baton with a 200 candle-power light. The entire parade, consisting of 350 people, carried a total of some 300 lamps.

The first and last parts of the parade passed without incident. The powerful light penetrated into every nook and cranny of the streets through which it passed. But suddenly, in the middle of the parade, all the lights went out. This mishap was caused by dirt from the water-tanks which fouled the pipe leading to the pump. Once this pipe had been cleared everything again went off as planned.

For two hours, the parade wended its way through the main streets of New York, attracting large crowds wherever it went. It was said that Edison himself was among the occupants of one of the carriages following the machines, and that he was loudly cheered by the onlookers.

Our readers will be interested to know that the presidential candidate who was recommended by this highly original parade was not the one who eventually received the majority of votes.

———

Electric home and street lighting by arc-lamps [1879]

A TRAVELLING ELECTRIC STAIRWAY LAMP

Monsieur Armand Murat of Paris has succeeded in devising an incandescent electric stairway lamp to light the way of those who come home or depart at a late hour after the gaslight has gone out. There at the foot of the stairs glows the lamp, and the latecomer need only lift the weight which is connected to it by a cord to find that it will rise upstairs before him as he mounts the steps until he arrives at his destination. Let him but release the weight and the lamp will go back downstairs of its own accord, ready to serve the next person who arrives. If one wishes to descend the stairs, the lamp can be pulled up in the twinkling of an eye by means of a chain which enables it to travel up and down. The lamp arrives, the counterweight is grasped, and now the light follows after the departing guest as he proceeds on his way downstairs. The lamp moves between two cables which constitute the conductors connecting it to an accumulator. According to Monsieur Murat, the lamp costs only some five centimes per night to maintain. The lamp is extremely simple to use but the technical construction of the system as a whole is fairly complicated, employing various pulleys, rollers, balls with holes drilled through them and a counterweight.

[1895]

Exhibition of the Sawyer electric light [1880]

CLOUD ADVERTISING

The projection of advertising messages on the clouds has been a much-discussed topic in recent years, but it is of course in America that this idea has been put into practical application for the first time.

During the last days of the Chicago Exhibition, a projector installed on the roof of the Palace of Arts and Manufactures, 200 feet above ground-level, informed the public every evening about the number of visitors attending the Exhibition that day, meantime entertaining them with the projection of words and drawings on the clouds.

When the Exhibition closed its doors, the equipment was taken to New York where it has kept the nocturnal pedestrians' eyes glued to the sky—provided it is overcast with clouds.

Mounted on top of the Pulitzer Building,

which houses the editorial offices of the *New York World*, the installation consists of a fixed-focus arc-lamp and mirror 2 feet 6 inches in diameter, producing powerful, parallel rays of light. A condenser fitted in front of the arc-lamp can be brought up from below by means of a chain and hand-wheel.

The drawing to be shewn is cut out from a piece of cardboard and interposed between the two lenses of the condenser. The equipment is mounted on a pivoting pedestal so that it can be aimed at a suitable cloud, following its every movement. The arc-lamp is intended for a current of 150 ampères and, since its voltage is 110 volts, it consumes 16·5 kilowatts, resulting in an operating cost of 15s. 6d. per hour in electricity consumption alone.

Should the necessary clouds be absent,

Cloud advertising in Paris [1894]

there is always the possibility of creating an artificial one by blowing steam into the air or sending smoke-producing rockets aloft.

Our engraving shows the device and the effect it produces on a cloud, in this case with an advertisement for the French scientific journal *La Nature*. This novel means of publicity is likely to be highly successful. But let us wait until it attracts the attention of the great industrialists.

ELECTRIC SEARCHLIGHT

A highly mobile searchlight to be used in warfare has been developed in France. Incorporating its own source of electric power, it consists of a steam-engine, a Gramme type dynamo, and an electric arc-lamp the light of which is turned into a beam by a concave mirror. It can be moved in any direction, and may be drawn by a single horse. Supplied in various sizes and candle-powers, the searchlight clearly illuminates objects at a distance of one mile when fitted with a mirror one foot in diameter; the use of

Electric lighting equipment for battlefield warfare [1886]

ELECTRICAL JEWELS

Monsieur G. Trouvé, of Paris, has conceived the idea of manufacturing artificial jewellery which, although entirely produced by craftsmanship, will in one respect by far surpass natural diamonds. These cannot, of course, display their brilliance unless they receive light from the outside: a diamond cannot sparkle in the dark, at most preserving a faint afterglow, whereas Trouvé's artificial jewels have an interior light source of their own from which will emanate the most brilliant, coloured rays of light, even in the dark.

A few of these artificial gems are shown in our engraving. They consist of a hollow enclosed in red and white glass lenses with ground facets and are not very different in shape from polished rubies and diamonds. Appearing as buttons on hair-pins, tie-pins and scarf-pins and even as knobs for walking-canes, they are intended especially to be worn as an adornment by stage artists and dancers.

The space inside the jewel contains a tiny electric light bulb connected by means of two thin conducting-wires to a small galvanic battery that may be concealed between the pins, in the hair or in one's garments. A battery giving off enough electricity to light the bulb for half an hour weighs less than 12 ounces.

There are also bouquets of artificial flowers which, when worn in the button-hole, will radiate electric light by means of, say, electric bulbs confined in a porcelain mould resembling a snowdrop. The electric circuit is closed by reversing a glass tube filled with mercury. The sudden illumination of such a nosegay will engender genuine surprise.

These trinkets will surely win great popularity, especially in the cities where ample facilities exist for charging small accumulators.

Electric jewels [1884]

Trouvé's electric jewels [1884]

a 3-foot mirror and carbon electrodes of double thickness extends the range to 4 miles.

For coastal defence purposes, the searchlight is useful for detecting the approach of enemy warships which thus will be an easy target to the shore batteries. In fortresses, the searchlight will prove invaluable for lighting the surrounding grounds, preventing the enemy from reinforcing his positions under the cover of darkness. [1874]

Photography by electric light [1879]

than a penny in his pocket. By inserting this coin into one of the ubiquitous automatic machines which one encounters everywhere it is veritable child's play to become the owner of one of these fine things.

And now an apparatus has also been invented which will permit us to enjoy electric light for fully half an hour, and that on payment of a negligible sum. If a penny is inserted in the slot marked A it will fall into B and then one can depress the knob which bears the legend 'Push hard'. This movement winds up a clockwork motor which connects up a circuit of several accumulators and a lamp for the space of half an hour. If the device is out of order—if, for example, a filament is broken—the coin will tumble out again at C.

The object of the inventors is to give travellers in trains and aboard ships a cheap but abundant supply of soft light for a limited time whenever they require it, thus enabling them to read, write, play games or pursue any of the hundred and one activities with which we can while away the time when on a journey. In this respect the electric incandescent lamp puts its humble sister in the shade—we refer, of course, to the gas lamp which in our third class carriages scarcely affords sufficient light to light a cigar by.   [1890]

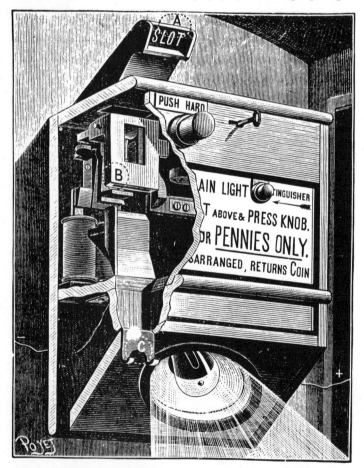

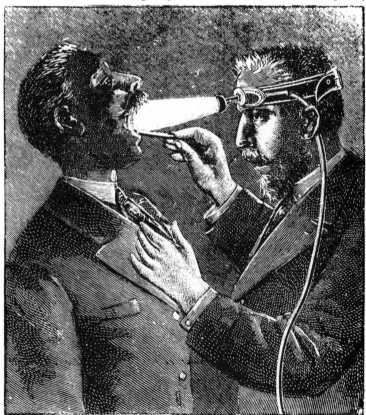

Laryngoscope with electric light [1897]

ELECTRIC LIGHT FOR A PENNY

Nowadays anyone who wishes to know his weight, have his portrait taken, his future foretold or who wants a travel insurance, a savings certificate, a newspaper, a packet of chocolate, or sweets, or even a squirt of perfume on his handkerchief, need have no more

ELECTRIC LARYNGOSCOPE

For closely examining the cavities of mouth and pharynx, the electric laryngoscope will prove useful. Consisting of a small electric bulb and a concave mirror, the instrument is strapped to the physician's forehead, leaving his hands free for the examination.

TESLA'S EXPERIMENTS WITH ALTERNATING HIGH VOLTAGE CURRENTS

While in Europe alternating currents with frequencies not exceeding 100 per second were still being studied eagerly for practical applications, reports were received from America in 1891 that most surprising experiments were being carried out there with alternating currents of 15,000 cycles. The initiator of these studies was a Hungarian employed with the Westinghouse Company—Nicola Tesla. With remarkable talent he has conducted experiments and research in a hitherto almost unexplored field: that of alternating currents of extremely high voltage and frequency. He gave an account of his work before the American Institution of Electrical Engineers in New York in a lecture which has since become famous. It made an indelible impression upon the audience, both on account of the brilliant experiments and the completely new vistas it has opened. His work places Tesla among the greatest of our present-day scientists and inventors such as Edison, Graham Bell and Thomson.

When the news of Tesla's experiments reached Europe, he was approached by the most prominent scientific circles in Britain and France who invited him to repeat his experiments in those countries. These lectures were attended by large and enthusiastic audiences which included men of great authority in the fields of the theoretical and applied sciences. After three hours of lecturing to an enthralled and fascinated audience, Tesla was compelled to admit that he had discussed only part of his research work.

Tesla uses two different types of equipment for generating his alternating high-frequency currents. One is a dynamo with 384 wire coils and an equal number of field magnets rotating at 50 revolutions per minute, thus producing an alternating current of $50 \times 384 = 19,200$ cycles per second. Tesla also uses a special type of transformer. Its primary coil has only a few windings and is connected in series with a spark-gap, a condenser and the secondary winding of a Ruhmkorff-type induction-coil. With the second combination, tensions of *half a million volts* and scores of thousands of cycles per second can be generated, producing most impressive discharge phenomena in the open air and in glass tubes filled with rarefied air.

In the air, these currents engender electrical fireworks of unprecedented splendour which assume the weirdest shapes, forming luminous fans and plumes of gossamer-like texture. Amazingly enough, these ultra-high voltages are in no way dangerous, thanks to their high frequency. In Berlin, Tesla placed himself between two of his assistants who were almost 15 feet apart, each of them touched one pole of the high-voltage transformer, and when Tesla reached out to them with his two arms, wavy bundles of violet-coloured electric fire shot forth from his fingertips, spreading out to one assistant's hand and to the other's forehead. This to the great dismay of some of the spectators, until they noticed that the experiment was harmless and painless!

One of Tesla's most striking experiments was his demonstration with the 3-foot-long Geissler tubes. For that purpose, two metal bars, 10 feet in length, attached to the floor and ceiling, were connected to the poles of his high-voltage transformer. When Tesla moved two Geissler tubes into the field between the two bars, they became luminous over their entire length without being connected either to the metal bars or to the transformer. In the words of one reporter: 'Tesla stood there

'Tesla stood there like the archangel, brandishing the flaming sword!' [1893]

like the archangel, brandishing the flaming sword!'

Tesla claims that in the future such glass tubes filled with rarefied air and placed in a field of alternating current will be an ideal source of light, because in comparison to the incandescent lamp these tubes produce very little heat so that almost all the electrical power is converted into light. While in America the use of alternating currents of only a few hundreds of volts has led to fatal accidents, and electricity is even employed for capital punishment, it is amazing to note that Tesla's extremely high voltages do not constitute any hazard to human life. 'I felt,' Tesla said when he was about to try them out for the first time, 'as if I were poised to jump from Brooklyn Bridge.'

From his numerous experiments, Tesla concludes that, in future, a method may be developed to channel the powers found throughout nature, and to extract them directly from the medium which carries them. 'Everything around us,' he says, 'is moving; a means must be found of harnessing this kinetic energy directly.' He further believes that through his experiments, ways and means will be discovered to convert electricity and other natural powers into radiation without the intervention of conductors—this he regards as more than just a vision.

Time alone will tell whether Tesla's ideas are correct. None the less this need not prevent us from voicing our admiration for his experimental talent, and joining in the recognition accorded to his merits by the most authoritative scientists, both in the Old and New World.

Recent Tesla experiments in Berlin [1894]

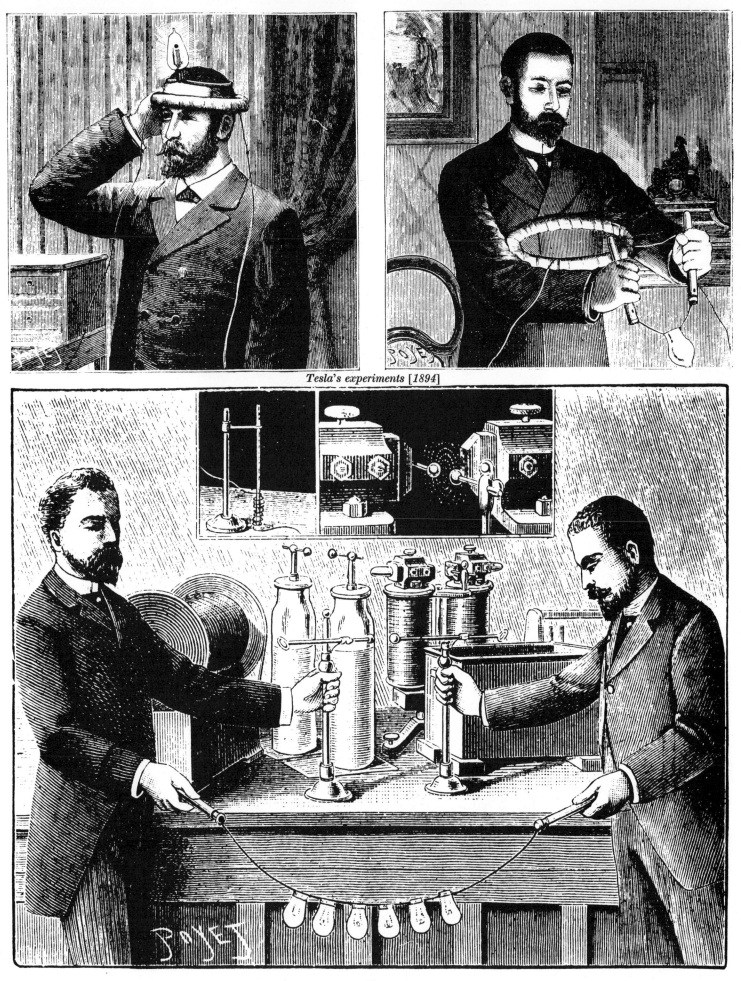

Tesla's experiments [1894]

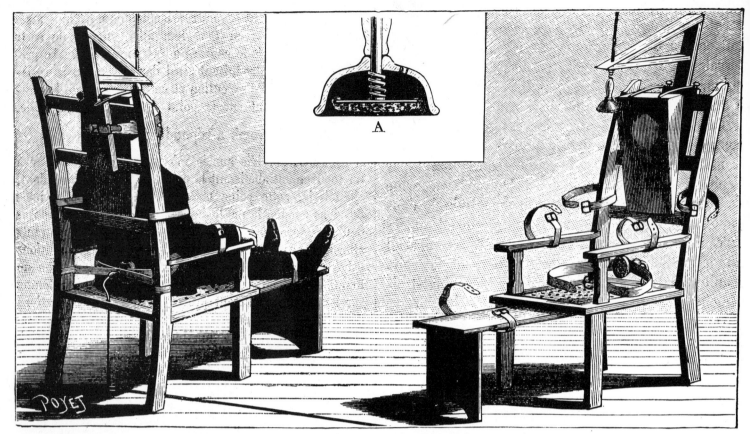

Front and rear view of the electric chair to be used for Kemmler's execution. A: the electrode to press against the head [1890]

ELECTRIC CHAIR

The State of New York may justly congratulate itself on the fact that the barbaric punishment of death by hanging is to be abolished in favour of a more humane and scientific method of execution: as from 1st January 1889, criminals will be put to death by electrocution. The engraving gives an impression of what the 'electric chair' will probably look like. The poles of a dynamo are connected by a switching device to a metal electrode clamped round the condemned man's head, and to the metal seat of the chair, sponges or wet cloths being applied at the points of contact to ensure a perfect electrical connection. Extensive experiments carried out with dogs have shewn that electrocution causes almost instantaneous death, eliminating the gruesome writhing movements of the hanged in the moments before death ensues. There is no doubt that for a civilised country which wishes to put an end to the barbaric horrors of the past the electric chair represents the best method of inflicting the death penalty.

ELECTROPLATING THE DEAD

Dr Varlot, a surgeon in a major hospital in Paris, has developed a method of covering the body of a deceased person with a layer of metal in order to preserve it for eternity. The drawing illustrates how this is done with the cadaver of a child. The body is first made electrically conductive by atomising nitrate of silver on to it. To free the silver in this solution, the object is placed under a glass dome from which the air is evacuated and exposed to the vapours of white phosphorus dissolved in carbon disulphide. Having been made conductive, the body is immersed in a galvanic bath of sulphate of copper, thus causing a 1 millimetre thick layer of metallic copper to be deposited on the skin. The result is a brilliant red copper finish of exceptional strength and durability.

Switch gear for the electric chair [1890]

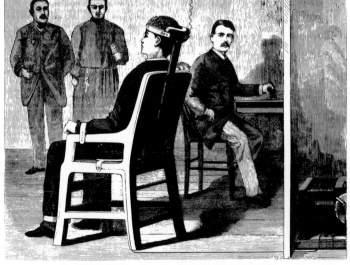

First design for the electric chair [1890]

Electroplating the dead [1891]

Demonstration of the Wimshurst high voltage electrical machine in Paris [1894]

Equipment for generating high-tension alternating current for experimenting with electrical discharges in glass tubes containing rarefied gases, such as Geissler tubes [1898]

PROF. RÖNTGEN'S X-RAYS

In December 1895, Dr W. C. Röntgen, a professor in the Imperial University of Würzburg, Germany, made a preliminary announcement before the Physio-Medical Society of that city about a 'new type of rays.' Dr Röntgen started his lecture as follows:

'If the discharges of a fairly large Ruhmkorff induction coil are allowed to pass through a Hittorf-type vacuum tube or through a Lenard, Crookes or other similar apparatus in which the air has been suitably rarefied and if the tube is covered with a snugly fitting mantle of thin, black cardboard, it has been found that a screen covered with barium platino-cyanide brought near the apparatus in a completely dark room will light up brightly, or fluoresce, at each discharge. This phenomenon is noted even at a distance of 2 metres from the apparatus.

'The most striking feature about this phenomenon is that an amount of radiation passes through the black cardboard cylinder, which is impervious to visible or ultra-violet rays generated by sun or arc-lamp, and that this radiation is capable of producing vivid

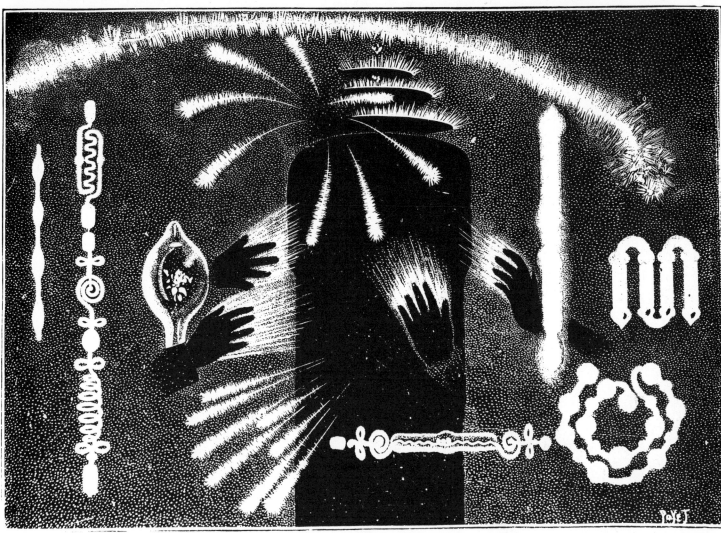

Luminous electrical discharges of high-tension alternating currents in the air and in Geissler's tubes [1898]

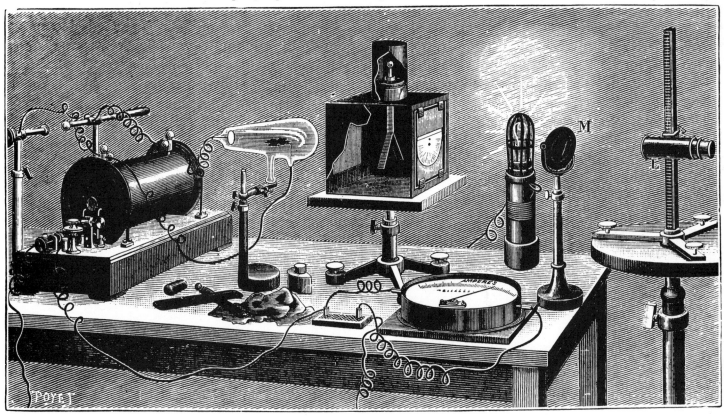

Apparatus for experiments with electrical discharges producing light effects and X-rays [1896]

Edison's surgeon's X-ray apparatus [1896]

pages. A similar observation was made with two decks of whist-cards. The rays also pass through thick logs of wood. To indicate that their nature is unknown, I should like to call the newly discovered rays 'X-rays'. The fluorescence can still be seen behind plates made from copper, silver, lead, gold and platinum provided they are not too thick. A

fluorescence; we shall first investigate whether or not other bodies also possess these same qualities. It will soon be found that all bodies allow these rays to pass through them, albeit to a widely varying degree. Here are a few examples. Paper is highly transparent to the rays: I saw the fluorescent screen clearly light up behind a bound volume of about 1,000

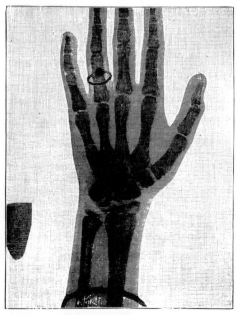

X-ray photograph of a girl's hand [1896]

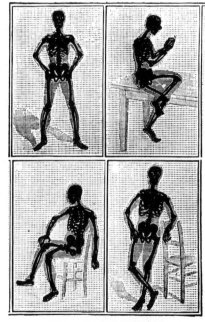

X-ray fantasies [1896]

lead plate 1·5 millimetres in thickness is impenetrable, however. In more than one respect it is important to note that photographic plates have been found to be sensitive to X-rays.'

These, then, were some extracts taken from Professor Röntgen's first announcement of his invention.

When the inventor permitted the rays to penetrate through a hand on to a photographic plate underneath, development revealed that the bones of the hand were clearly visible on it. The importance of these X-rays to medicine was instantly realised. In the few months that have elapsed since the discovery was made, it has been found possible to trace many metal objects such as bullets in the human body with the aid of X-rays, so that they could be removed by operation. For the first time, too, doctors have been able to obtain a clear picture of complicated leg fractures. Even now it can be stated with certainty that Professor Röntgen's invention will prove to be a boon to humanity.

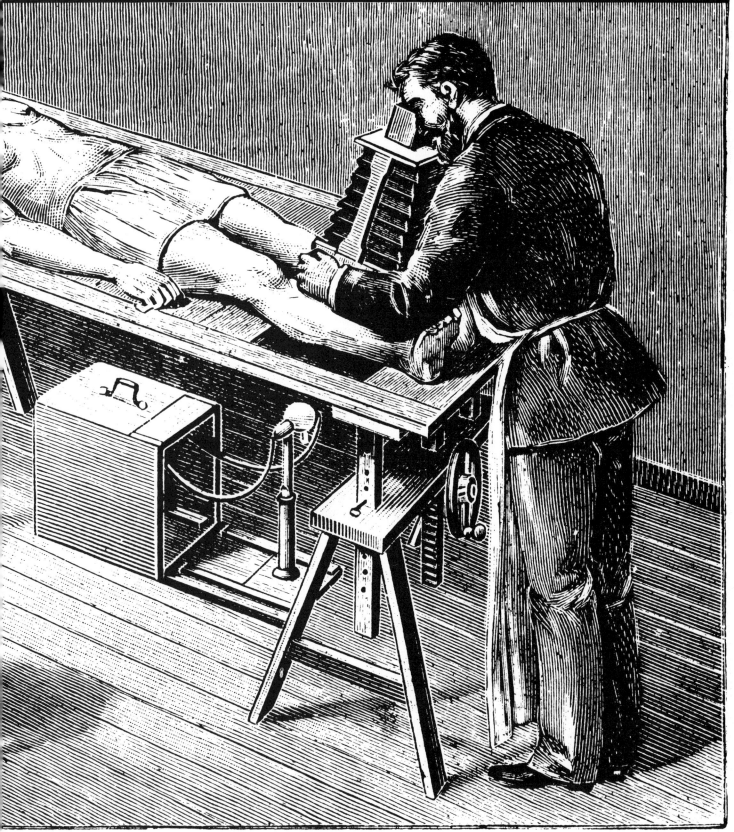

Equipment for radioscopic examination and photography [*1897*]

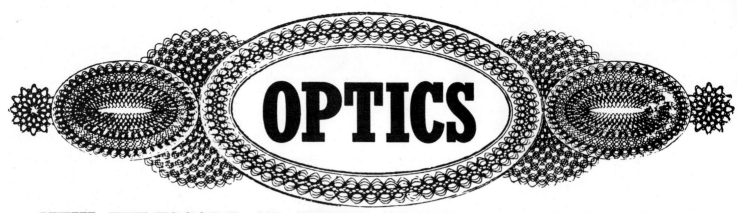
NEW TELESCOPE OF SHORT FOCUS

We represent a new telescope devised by Monsieur Leon Jobert, the able Director of the Popular Observatory at the Trocadéro, Paris.

This instrument is like the Cassegrainian telescope in form, and is of short focus, its parabolic reflector being only half the focal length of those of Foucault. It is of variable latitude, or, in other words, may serve for all points of the globe. In order that the observer may, without changing his position, be able to sweep the whole heavens above the horizon, the ocular is located at the intersection of the polar axis with the axis of declination. The sides of the tube are furnished with two supports, which are jointed round the horary axis, and pass through two other large supports that form a part of the last-named axis, and that are connected with each other by a turned circle moving over two large rollers. The body of the telescope is balanced by two weights whose supports are fastened to the axis of declination.

By means of a hand-wheel the instrument may be fixed at the latitude of the locality where it happens to be placed, in such a way that the prolonged polar axis is parallel with the axis of the earth and points to the celestial pole. The instrument is furnished with a polar circle and a circle of declination with verniers that are moved by endless screws. In the figure the observer is represented with his hand on the hand-wheel which actuates at the operator's will, either rapidly or very slowly, the axis of declination. A clockwork movement is transmitted by bevel wheels and an axle, to a wheel which revolves loosely on the axis of latitude formed by the bearings of the large arc; and from this point motion is transmitted to the axis of the endless screw, and thence to the endless

New telescope of short focus [1878]

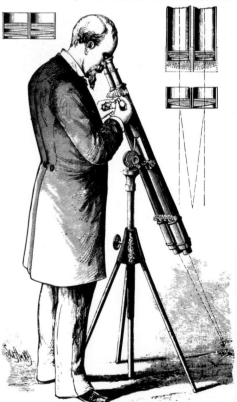

Binocular telescope with microscopic objectives [1880]

The new planetarium in Paris [1880]

NEW AND GIGANTIC TELESCOPE

screw which actuates the polar axis. With this instrument the observer can sweep every point in the heavens without changing his position, the only change he makes in the latter being that of moving with the instrument, which makes one complete *La Nature* revolution every twenty-four hours.

Among the many ideas which have been elicited by the discussion in these columns [*Scientific American*] regarding a gigantic or 'million dollar' telescope, we have recently had submitted to our examination one which seems to us quite novel, ingenious and,

although untried, not unpractical. It is a scheme for a huge instrument, to be built on either the Gregorian or Cassegrainian system, in which the image is first received on a large parabolic mirror located in a position diametrically opposite to the objective in a

107

refracting telescope, thence reflected back to a secondary mirror which, in accordance with the respective systems, is either concave or convex, and by the last re-reflected to the eyepiece, the tube of which passes through an orifice in the centre of the large glass. It is hardly requisite to explain the immense labour and almost insuperable difficulties which would be encountered in constructing a reflector of the proposed size—10 or 15 feet in diameter—of metal, and mounting the same. The great mirror in the telescope in Melbourne, Australia, though but 3·8 feet in diameter and weighing 3,498 pounds, required 1,270 hours of continuous labour to bring it into the last polishing stage, while its adjustment and mounting exacted the nicest engineering skill. In brief, it may be safely asserted that a metallic mirror, of the large size above noted, supposing it could be successfully constructed, would, from its great weight but far more on account of its consequent flexure, be practically useless.

Mr Daniel C. Chapman of New York, who is the originator of the plan, suggests both a mode of making a mirror of light weight and also a method of supporting the same. The

The new telescope of the observatory in Paris [1876]

New and gigantic telescope, designed by Daniel C. Chapman [1874]

reflector, he says, may be constructed of glass. A mould of clay, metal, or cement, of the required shape, is carefully formed and placed in a suitable furnace, cavity upwards. Over the latter a huge plate of glass is disposed, and the heat applied. At a certain temperature the glass begins to soften, and in such state may be bent, fitted into the mould, and subsequently annealed. The whole is then removed and placed on a plane. The glass is taken from its bed, disposed convex side up, and a backing of cement or plaster, the composition of which is previously determined by experiment so that it shall have

the same co-efficient of expansion as the glass, is applied to several inches of thickness, The mirror is next inverted, placed on a turning table and carefully ground or finished within, into the exact form necessary. But little labour, comparatively speaking, will here be required, as an approximate or very nearly true curve will, it is believed, be taken by the glass in fitting itself to the mould. The reflecting face is, lastly, silvered by Dr Draper's process, a solution of Rochelle salts and nitrate of silver being applied, which very quickly deposits a fine uniform metallic surface. It will be noted that the inventor

thus obtains a reflector of light plaster and glass, the weight of which is quite small. On the rear of the plaster backing are made a number of projections, arranged with sockets to receive the end of any number of braces. The latter are of wood, strong and well seasoned, and covered with some preserving material. These, extended from various points on the back, meet at the centre of a huge copper sphere, which encloses the entire apparatus except the mirror, and then, intersecting, spread again to abut against the interior periphery of the globe.

––––––––

THE GIANT TELESCOPE AT THE 1900 WORLD EXHIBITION IN PARIS

The usefulness of a world exhibition lies partly in the incentive that such an event induces to produce something hitherto unknown in the fields of industry, the arts, or sciences. Thus a few amateurs of the science of astronomy have had a giant telescope constructed for the forthcoming Paris Exhibition—an instrument which surpasses all existing telescopes. While the telescope of the Yerkes Observatory in Chicago (so far the largest in the world) has an objective 3 feet in diameter and a focal distance of 67 feet, the new telescope has an objective with a diameter of 4 feet and a focal distance of no less than 200 feet.

If a telescope of that size were to be mounted in the same manner as smaller telescopes it would give rise to many almost insurmountable problems. The dome under which the telescope would have to be accommodated would require a diameter of 215 feet, and it would have to be in constant motion to keep the objective facing the aperture in the dome at a speed of 10 miles per hour. To eliminate this and many other drawbacks such as the fatigue to which the astronomical observer is subjected, it was decided to install Foucault's 'siderostat' instead of the conventional arrangement.

The siderostat is an instrument which reflects a portion of the sky in a fixed direction regardless of the diurnal motion of the heavens. It consists of a flat mirror which, placed before the objective and moved by a clockwork motor, always throws the rays of light into the fixed telescope from the same horizontal direction.

Made of $\frac{1}{8}$-inch-thick steel, the telescope tube has the enormous length of 200 feet. The objectives, weighing 2,000 pounds in their mountings—one for direct observation with the eye, the other for photographic exposures —are fitted on a truck on rails. Together with its mounting, the plane mirror with a diameter of nearly 7 feet weighs 14,750 pounds. It took eight months to cast, grind and polish it, and

The new meridian telescope in the Paris observatory [1877]

The giant telescope of the 1900 World Exhibition in Paris. 1, total view

The great equatorial telescope of the Paris observatory [1883]

this was done to the close tolerance of 0·00025 inch.

The giant telescope has a magnifying power of 10,000 which would permit an observer on earth to follow the manœuvres of an army group or the movement of a transatlantic steamer on the moon.

As shewn in the engraving, this telescope, the largest in the world, will be used to project on a screen not only the sun, but also the moon and possibly a few other planets, as well as some nebulae and binary stars.

2, the siderostat: 3, the objective part: 4, the ocular part of the telescope [1899]

A new pocket size photographic camera [1875]

The photographic hat [1887]

THE PHOTOGRAPHIC HAT

In the search for small, easily portable photographic apparatus various devices have been invented. A typical example is the recently invented photographic hat which permits the wearer to take photographs without being noticed. That this is indeed possible appears from a notice received from Mr O. Campo of Brussels:

'When staying in a seaside resort in August 1886,' says Mr Campo, 'I received a letter one day and, upon opening it, found to my amazement that it contained three photographs in which I had no difficulty in recognising myself. But what photographs they were! Certainly not such as might be calculated to tickle my vanity. In the first photograph, I was shown at the very moment of entering the water, and my face reflected all too clearly the sensation of the first contact with the cold sea-water. Really, one would not approach a lady with such gestures on the shore.

'The second picture had been taken while I was blowing out a mouthful of water which I had involuntarily gulped in, while my facial expression made it obvious that my taste buds had been stimulated in a far from pleasant way. In the third picture, I resembled a bedraggled poodle rather than a civilised man. I emerged from the sea, dripping wet.

'It cannot be denied that the three exposures were a true reflection of what had happened two days before. On investigation, I found that a good friend had availed himself of the opportunity of making several photographs of me while I was bathing, and had done so with the aid of a *photographic hat*.'

First patented in 1886 by the Luders brothers at Görlitz, Germany, the photographic hat inspired *Punch* to produce the following outpouring in rhyme:

If they knew what I wear when I walk in
   the street,
It should be quite a terror to people I meet;
They would fly when they saw me, and ne'er
   stop to chat,
For I carry a camera up in my hat.

A Herr Luders, of Görlitz, has patented this,
And I think the idea is by no means amiss;
With a hole in my hat for the lens to peep
   through,
And a dry plate behind, I take portrait or
   view.

How to use the roll-film camera [1895]

Photographing by the magnesium light [1884]

A magnesium torch [1895]

Should I meet, when I chance to be taking the air,
With a lady who looks so surpassingly fair;
If I wish to preserve her sweet face by the sun,
Why I just pull a string, and the photograph's done.

A photographic studio [1888]

113

Parsell's portable photographic camera and tripod [1885]

I admire, say, a sea-scape, or else chance to
  look,
With the eye of an artist, on picturesque
  nook;
There are plates in my hat, if I poise it with
  skill,
That will take any beautiful view at my will.

A photographic tricycle for tourists [1885]

If I'm stopped in the street—that may happen, you know—
By a robber whose manners are not *comme il faut*,
His identification should never be hard,
There's my neat little photograph in Scotland Yard.

So we'll all wear the hat made by science complete,
With a camera, lens, and a dry-plate *en suite*;
And take views in the street with its bustle and traffic,
With the aid of this German's strange hat photographic.

Photographic tricks [1889]

115

PHOTOGRAPHIC RIFLE

Professor E. J. Marey in Paris, who is engaged on the study of the way in which birds move, has hit upon the idea of making a kind of photographic machine-gun with which a series of pictures of a bird in flight can be obtained within a very short space of time. The difficulty here was not the sensitivity of the bromide of silver and gelatine layer on which the pictures must be formed, but in the speed with which the sensitive plate must move in order to come into the focal spot of the lens. Marey succeeded in constructing a device the size of a hunting-rifle which photographs the object aimed at twelve times in one second, each picture requiring a pose of only $\frac{1}{720}$ second.

The barrel of the rifle is a tube containing the camera lens. At the rear there is a cylindrical drum attached to the butt of the rifle and containing a clockwork motor. The system of gear-wheels which imparts the necessary speed to the various parts is set in motion by pulling the trigger. These parts are attached to a shaft which rotates twelve times per second. First, there is a metal disc containing a tiny window which permits the light from the lens to enter twelve times per second, and for $\frac{1}{720}$ second each time. Behind this is a second metal disc which has twelve apertures, against which the sensitised glass plate is placed. This second disc and the glass plate rotate only once per second, stopping briefly after each rotation so that the image of the bird can enter through the window in the first metal disc twelve times in succession and fall on different parts of the glass plate.

After some aiming practice, Marey obtained very satisfactory photographs in which each complete beat of a seagull's wing was depicted in three exposures. Marey considered this inadequate and doubled the speed at which the glass plate and the metal discs rotated. In this way he obtained very good pictures, despite the fact that the light impression of the bird's image then struck the sensitive silver layer for only $\frac{1}{1440}$ second.

Marey is now building up a large collection of various species of birds in flight, both hovering and flapping their wings, and in differing conditions of wind direction and velocity, varying from absolute calm to storm force. He has even succeeded in photographing bats despite the lateness of the hour at which they fly and the unpredictable nature of their flight. He hopes that a close study of these pictures will help him to cast a new light on the way in which winged creatures are able to rise into the air and fly. This may also prove useful with respect to the—so far consistently unsuccessful—attempts made by man to create flying-machines.

The photographic rifle [1882]

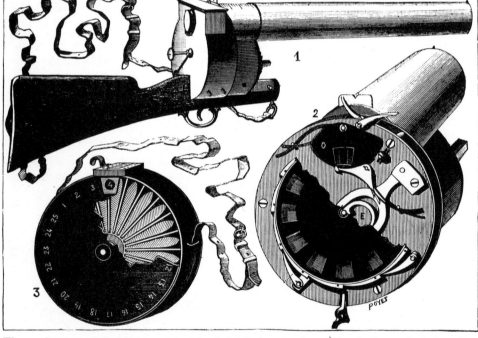

The mechanism of the photographic rifle. 1, total view: 2, view of the shutter and window-disc: 3, magazine containing 25 photographic plates [1882]

Photographs of a gull in flight [1882]

INSTANTANEOUS PHOTOGRAPHS OF MEN AND ANIMALS IN MOTION

The photographer Muybridge of San Francisco has succeeded in taking photographs of fast-moving animals and people for the purpose of making a thorough study of the various positions of their bodies and limbs while in motion.

A small racing-track has been laid out, 150 feet long and 55 feet wide. Throughout the length of the site, backgrounds have been set up, one side being black and the other white, and these can be reversed at will.

Opposite the background is mounted a set of 24 cameras, each having a lens 32 millimetres in diameter and a focal distance of only 9 centimetres. In addition, a battery of 12 cameras has been arranged at either end of the racing-track to take head-on photographs of the people and animals on the track, while the other cameras photograph them in profile. This brings the total number of cameras in operation up to 48.

Initially wires were strung tautly across the floor and these were broken one after the other by the man or animal moving over the track. On breaking, each of these wires closes an electrical circuit, thus causing the objective of the camera concerned to be opened for a brief moment. The engraving shows the shed where 24 photographic cameras are placed.

An entirely new process has been used for the required photographic plates, replacing the usual substance, collodion, with a solu-

The racing-track used by Muybridge for his instantaneous photographs [1882]

does not fail to make its appearance as well.

Muybridge has made fifteen successive photographs of a single beat of the wing of a bird in flight. Studied under a magnifying glass, these pictures prove that every single quill performs movements which are quite independent of the other. In the photographs made of racing-horses one may observe the most singular movements of the legs. At times, the horse seems to float through the air; one moment, all the legs appear to be bunched up underneath the animal's belly—the next, a front leg and a hind leg are stretched out so unnaturally that, but for the truthfulness of the photograph, this position would be regarded as impossible.

With the short exposures of up to $\frac{1}{5000}$ second, Muybridge's photographs have come a long way since his first efforts which were no more than silhouettes. By using an abundance of light obtained, among other things, by reflecting the sun's rays from white or mirror screens, by steadily improving his lenses and by exercising great care in preparing the exposures, Muybridge has succeeded in making pictures showing a hitherto unknown richness of detail. His work is of paramount importance both for the arts and for the sciences.

———

tion of gelatine and water containing finely dispersed silver bromide. Spread out over a glass plate and dried, such a solution forms a thin film which is so sensitive to light that a photograph can be obtained in less than $\frac{1}{500}$ second, a period of time which is so short that even a running horse cannot be seen to move forward in so brief an interval. However, Muybridge has managed to achieve even shorter exposures of less than $\frac{1}{2000}$ second or even $\frac{1}{5000}$ second!

Muybridge has ceased to use breaking wires to open the camera objectives by electricity. Instead he now employs a chronographic clockwork which records the exposure time of the plates by means of the vibrations of a tuning-fork. The clockwork can be adjusted to ensure that 24 plates are successively exposed at intervals which can be varied.

The photographs obtained by this method are of such eminent scientific value that the University of Pennsylvania has taken over the supervision of the work and has financed a large share of the cost to the tune of no less than $30,000. The remainder is covered by the proceeds from the publication of the book *Animal Locomotion*, the subscription price for which is $100; about 150 persons have so far availed themselves of this offer.

That Muybridge's effort has indeed been considerable, is proven by the fact that he has exposed well over 100,000 plates while his work *Animal Locomotion*, printed in photo-type, contains pictures of animals and men in more than 20,000 positions on 781 foolscap pages. Where human beings are concerned, Muybridge has literally 'frozen' all sorts of movements involving the carrying out of work by men and women and the playing of games and sports by people of every age, whether naked or clothed. Children climbing stairs, women and men performing gymnastics, both the healthy and the sick he has captured one and all in the infinite variety of movement. In addition, he has photographed every animal in motion he could get before his cameras; this includes not only horses and dogs, etc., but also other suitable animals which he found in the zoo, so that where the pigeon is present, the eagle

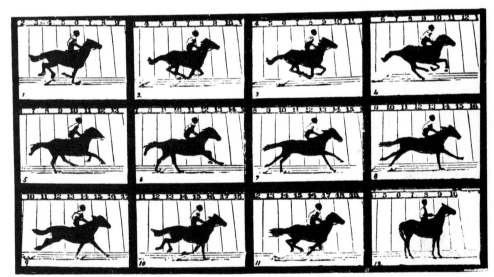

Muybridge's instantaneous photographs of galloping horses [1882]

THE ELECTRICAL TACHYSCOPE AND THE KINETOSCOPE

The famous photographer Anschütz has succeeded in constructing a device with which he can project his splendid snapshots of various living creatures, recorded chrono-photographically, on a screen. This device, called the 'rapid viewer' or 'tachyscope,' is constructed as follows. An iron wheel, mounted on an iron base reposing on a mobile carriage, can be made to rotate round an axis by means of a crank-handle. A series of little windows is fitted to the outer rim and on these are placed pictures of the object in the various successive positions of its movement.

A small pin situated centrally below each figure makes momentary contact when the wheel is turned and thus provides an electric current generated by several accumulators. This current flows through the primary winding of a Ruhmkorff induction-coil. An induction current is thus set up in the secondary winding of this coil which, passing through a Geissler tube, causes the latter to light up brightly. This light occurs only at the moment when each picture is travelling past the eyes of the onlookers. The impression of each figure lingers until the following figure

The electrical tachyscope of Anschütz [1895]

appears and thus is created the illusion of seeing birds in flight, horses trotting along, and similar sights.

Anschütz has had great success in America with this device. He kept the whole apparatus concealed behind a wall. Our artist has drawn the scene with a part of this wall cut away in order to show us the spectators behind it.

Meanwhile, Edison too has developed a machine permitting the observation of chronophotographs. He calls it the 'kinetoscope.' In this system the negative of the chronophotographs are printed in positive form on a thin flexible strip of celluloid made sensitive to light and no less than 15 metres in length. On it are some 750 pictures, each 2 centimetres high, which can be observed

in the space of three-quarters of a minute—approximately seventeen pictures per second—through a slit in a rapidly rotating disc.

The chronophotographs are taken from life, the subjects being real actors, dancers, fighters, etc., preferably from amongst the most famous in their art. The most renowned theatre artists, champion boxers, wrestlers, lasso experts, Buffalo Bill sharpshooters, snake dancers, Eastern knife-throwers, Mexican duellists, kings and queens of the tightrope or trapeze—all have performed for Edison to be immortalised by the kinetoscope.

The apparatus itself consists of a wooden box into the top of which is fitted a viewing glass; looking down through this the observer sees the entertaining performance re-enacted within the box. No more than one person at a time can see the show. Assuredly, it is as yet no more than a toy, but of such a kind as grown-ups too delight to peek into. For example, you may see a Spanish dance, performed by a lady and gentleman with such exquisite expertise as will never have been seen before by many of us. On another occasion, three smiths are seen in a smithy busily hammering an iron rod on an anvil; when they grow tired, one of them seizes a bottle from which each in turn takes a few gulps of beer. Another time we witness a cockfight in which two cockerels go at each other hammer and tongs so that a flurry of feathers goes flying over the scene. When they grow tired, fellows armed with sticks goad them on so that the cockerels resume the struggle and continue until they are ready to drop. The onlooker sees everything proceed as naturally as in real life, although the whole scene on the celluloid strip occupies a space no greater than one-sixth of a visiting-card.

According to the latest reports, Edison is engaged on perfecting his invention so that, firstly, it will be able to reproduce life-size pictures of men and animals instead of

Interior of Edison's kinetoscope [1895]

Edison's first experiments of his kinetoscope in combination with his phonograph [1895]

Lilliputians and, secondly, will show pictures which talk or shout as well as move. For this purpose he will combine his new kinetoscope with his phonograph but this involves great difficulties with regard to achieving simultaneous reproduction of both instruments. There are also serious problems connected with the projection of the pictures, for when the images on the celluloid strip are magnified 500-fold the slightest discrepancy will spoil the performance.

Edison claims that by combining his improved kinetoscope and the phonograph he will be able to show a full opera, and states that the possibilities of this combination will be unlimited. We take the view that his expectations are somewhat exaggerated. To mention but one objection, to achieve a full impression of reality the photographic images would require to be capable of rendering objects in their own natural colours, and that is something which cannot as yet be done. If we may make so bold as to make a request of Edison it would be this: to solve the problem of colour photography. Then we should no longer need to hesitate over the answer to the question: What is Edison's greatest invention?

Reynaud's new projecting praxinoscope [1882]

THE PRAXINOSCOPE

By means of chronophotographic exposures as made by Muybridge in which the light was admitted at specific, regular intervals, the all-but-forgotten stroboscope can finally be converted into an instrument which will be useful for the arts and sciences. It will permit us to see a sequence of photographs of men and animals in motion as moving 'living' pictures.

In its simplest form, the new device is a 'wonder cylinder' or zeètrope. When its inside is covered with a series of photographs and the cylinder is rotated quickly, the spectator looking through the slits will see the figures in motion due to the fast sequence of impressions recorded by the retina of his eye. The various light impressions are combined and since they are photographs of the elements of a movement in their correct sequence, the individual pictures 'merge' into a flowing movement which is entirely realistic.

To make these moving figures visible to a large number of people at one time, the wonder cylinder can now be combined with the magic lantern. This has been achieved by Reynaud with his praxinoscope. The lamp of the magic lantern first throws its rays of light through the lens on to the screen, projecting, for example, some landscape or other; part of the light then passes through a prism which projects the light-rays on the moving figures. These reflect them back to the top lens, which throws them on the screen in magnified form, thus enabling the spectators to see the moving figures. Muybridge gave a few performances before The Royal Society in London in March 1882 with an apparatus of this kind. Not only did horses, dogs and bulls, etc. move across the screen at considerable speed, but the spectators also enjoyed the agility and skill displayed by gymnasts and pugilists.

Reynaud's praxinoscope for the home [1883]

STEREOSCOPIC PROJECTION

Bacon, the famous philosopher, claimed that man could see better with one eye than with two, his attention being more concentrated in that case.

'It is noted that, when we close one eye while looking in the mirror the pupil of the other eye is dilated.' 'However, when he was asked why we have two eyes, he replied: 'To have one left when the other is damaged.' Despite this explanation attributed to the celebrated philosopher, we believe that we have two eyes for better vision, and

Stereoscopic projection [1890]

other that seen by the left eye. Since that time, many more 'stereoscopes' have been presented to the public. However, they could be used by one person only.

This drawback is eliminated by the equipment recently demonstrated in Lyons, France. It consists of two slide projectors built into one housing, their projecting lenses being placed in tandem arrangement one above the other. The equipment makes it possible to superimpose two pictures of different colour on a screen. By taking two stereoscopic images of the same scene and giving them different colouring—for example red and green—and throwing these two slightly differing pictures on the same screen, they may be viewed by the audience in the proper perspective if means are adopted to ensure that the green image is, for instance, seen by the left eye of those present and the red image can be seen by the right eye only. For that purpose, the spectators were given spectacles with a red and green lens in order to eliminate the red picture from the field of vision of the left eye while their right eye was prevented from seeing the green picture because it was compelled to look through the green lens.

So far, the results have been encouraging, and there is little doubt that 'stereoscopic theatres' will before long be established in our major cities.

especially to see objects in their proper shape and perspective.

Although the effects of binocular vision were already known to the ancients, it was not until 1838 that Wheatstone, the British scientist, constructed the first stereoscopic apparatus, permitting a three-dimensional view to be obtained—by using both eyes—of two different images of an object, one having the perspective observed by the right eye, the

Reynaud's optical theatre [1893]

120

REYNAUD'S OPTICAL THEATRE

In 1892 Reynaud, the Frenchman who had succeeded in projecting the pictures of his praxinoscope on a screen, presented his Optical Theatre to the world. In this theatre it is not only a single sequence of movements which is shewn, but the illusion is extended to such a long series of effects that an entire act can be projected on the screen. This is not done by chronophotography but by means of images painted in various colours, i.e. made by hand on a transparent tape. The projectionist can move this tape in two directions by turning two cranks, thus causing the pictures to move through lantern B to be projected by lens C on an inclined, flat mirror M from which they are thrown on to the transparent screen E. A second lantern D projects the static and unchanging scenery on this screen and in this setting there appear the constantly changing figures that have been painted on tape A.

The sequence of pictures may be interrupted at any desired moment without interfering with the luminosity of the picture or its visibility on the screen. Consequently, the performance may also include pauses and repetitions which tend to increase the verisimilitude as well as the duration of the event. This permits entire pantomimes lasting from fifteen to twenty minutes to be shewn. The engraving represents an episode from a most amusing pantomime called *Le pauvre Pierrot* in which Harlequin, Columbine and Pierrot move like living statues.

For the rest, it is most difficult to draw a large number of positions in the proper sequence with such accuracy that the movements appear to be entirely natural.

THE CINEMATOGRAPH OF MESSRS AUGUSTE AND LOUIS LUMIERE

It is just over a year ago that we became acquainted with Edison's kinetoscope which enabled us to look through a spy-glass at 'living' photographs of deceptive reality. Since that time, every spectator has hoped that Edison would soon succeed in projecting such scenes on a full-size screen. Thus, a large audience would be able to see the moving pictures, possibly even in life-size. Although at that time it was stated that Edison had managed to produce such pictures, it is more than likely that the American performance was not entirely successful, for otherwise the American trumpets would surely by now have broadcast the news that the invention was ready to be converted into hard cash.

It now appears that the Lumière brothers of Lyons, France have stolen a march on the Americans in this field. In July 1895 they gave a performance with their invention, the cinematograph, which was warmly applauded by the audience composed of many scientists. Recently, we were present at such a show in the Kalver Straat in Amsterdam. Having paid 50 cents admisssion fee, the spectator sits down on one of the seats in front of a white screen, about 3 feet square, that has been pulled taut inside a timber framework. When the hall is full, the electric lamps are suddenly extinguished, and the transparent screen is seen clearly illuminated by transmitted light. Then one of the acts announced for the day is shewn. A picture with clear, sharp images, an act full of movement—not just resembling real life, but actually taken from real life.

To mention but one example, the factory of the Lumière brothers at Lyons is shewn at the hour when a large number of workers—men and women—are leaving the factory for their midday meal. They rush out into the street in groups, going in various directions, dodging passing cyclists and carriages,

Gaumont's pocket size cinematographic camera [1900]

chatting and jostling each other playfully, visibly glad to be free for this brief spell. The entire act lasts no longer than one minute, ending as suddenly as it has begun. The audience, genuinely impressed, voice their opinion in remarks such as "How wonderful!" and "What a pity it's finished already!" A second act follows almost immediately, and every performance includes about ten such acts, each more surprising than the next. Among the scenes that created widespread admiration was that of a small girl seated between her parents in the open air. Shewn in life-size, the girl was having her meal which she obviously enjoyed, while straightening her apron which was raised by the wind.

The audience were spellbound when they saw a train approaching from a distance, nearing the railway station until it had grown to almost full size and seeming to penetrate into the hall; they also enjoyed the view of the station itself with its hustle and bustle of travellers boarding and leaving the train.

Other scenes include children playing at the seaside, and cardplayers smoking and drinking in a pub. However, there is one general remark which applies to all of them:

indeed, we see real life, but it is silent—devoid of sounds or spoken words. Difficult though it may be, one further step remains to be taken: to make it possible to hear the players talking, the blacksmith hammering on the anvil, the cork popping out of the bottle, the fearsome roar of the stormy sea. Then the illusion will be perfect.

In order to permit an even larger audience to enjoy the performance, the screen is sometimes placed in a partition between two halls, the scenes being shewn in transmitted light in one of them and in reflected light in the other.   [*1897*]

The cinematograph at the World Exhibition in Paris [*1900*]

MICROSCOPIC DISPATCHES BY CARRIER PIGEON

Since the advent of the electric telegraph, carrier-pigeon post is all but dead and buried. And yet it was thanks to the use of the carrier-pigeon service that the Rothschild brothers amassed the immense fortune which they now possess. In 1815, carrier-pigeons brought to the houses of Rothschild in Paris and London the tidings of Napoleon's defeat at Waterloo. For three days, the fortunate bankers had time to buy at their leisure, and at the lowest conceivable price, a vast quantity of shares on the London Stock Exchange, until the moment when the outcome of the battle became known to the Government itself.

In addition, the intelligence was also transmitted by the British optical telegraphs, but it was interrupted by fog after the two words 'Wellington defeated' . . . which originally gave the impression that the English had been beaten by Napoleon, thus causing a sharp drop in the price of Government securities. In fact, the full message ran as follows: 'Wellington defeated the French at Waterloo'—a piece of information which was held exclusively by the house of Rothschild for three days.

The carrier-pigeon post was revived during the siege of Paris by the Prussians in 1870. The pigeons were taken from the besieged capital by balloon to the unoccupied part of France. There, dispatches were attached to the birds and they were released, whereupon they flew back to Paris at great speed. At first the dispatches were sent in minute handwriting on extremely thin slips of paper, but when more dispatches arrived than could be carried by the pigeons, a scientist named Dagron, found a solution with his method of employing microscopic membranes. By means of microscopic photography, the messages were transferred to the membranes on such a reduced scale that many hundreds of dispatches could be accommodated on a

Projection of the microscopic dispatches [1889]

single membrane 1¼ by 2 inches in size. The birds carried the membranes in tubes made of quill which were attached to the pigeons' tails. Thus, a single pigeon could transport thousands of dispatches although these birds cannot carry a weight of more than 1 gramme. As soon as the pigeons arrived in Paris at their respective lofts, the membranes were taken to a specially equipped reading-room where they were magnified, copied and sent to the addresses. The work of magnification was carried out as follows: in the middle of the room which had been darkened for the purpose, an apparatus for electric light—an arc-lamp—was connected to batteries placed below the table. A set of glass lenses, some of which had been ground and polished to a concave shape, was fastened to the front of the apparatus to direct the rays of light on to a screen. The dispatch, sandwiched between two small glasses, was placed between the light source and the lenses and in this way it was projected on the screen in legible size.

During the period when overland communications with Paris were interrupted, a total of 95,581 dispatches and all kinds of special telegrams were entrusted to the carrier-pigeon service, which thereby earned a total sum of 432,524 francs.

STEREORAMA AND TRANS-SIBERIAN RAILWAY PANORAMA

The Stereorama 'Poème de la Mer' [1900]

It will come as no surprise to our readers that the World Exhibition in Paris includes several panoramas. Since we have already become accustomed to panoramas at country fairs of any importance, it is small wonder that they are not lacking at the greatest fair ever held. An ingenious means has been devised to enliven the stationary panorama and hence to make the picture shewn more realistic by the artifice of movement. This is an elaboration of the well-known illusion of the person sitting in a stationary train who sees a train pass by and thinks that himself is moving.

In the Stéréorama or 'Poème de la Mer,' the spectators imagine themselves to be occupying the cabin of a steamer sailing along the Algerian coast from Bona to Oran. As the ship departs, the sun rises above a perfectly calm sea which, after Bougie, becomes choppy with froth-crested waves. Before Algiers the sun burns brightly in the sky. The panorama of the city with its white houses and domed minarets is outlined on the

the smoke rising from the funnels is particularly realistic in its effect.

Unlike the usual panoramas, the background is painted on the outer mantle of a slowly revolving cylinder with a wide protruding edge carrying forty concentric sheet-metal screens 4 inches in height on which the waves have been painted. These screens are moved up and down by an electric motor through a linkage system including rods, hinges and wheels. The illusion of reality is as perfect as it is gripping.

The second panorama, The Trans-Siberian Railway, has been built by the Compagnie Internationale des Wagons-Lits close by the Russian and Chinese exhibits. The spectators sit in real railway carriages of the Compagnie; there are only three carriages but each of them is 70 feet long and contains saloons, dining-rooms, smoking-rooms, bedrooms and dressing-rooms, bars, a kitchen equipped to satisfy the taste of connoisseurs, a well-furnished hairdressing parlour and even a bathroom with gymnasium. All fully and lavishly equipped for the comfort of the passengers. There is also space for spectators in front of the carriages. All the noteworthy things that would pass before their eyes on the 6,300-mile-long, fourteen-day trip from Moscow to Peking—once the railway is

The Stereorama: mounting the first screens representing the waves, and the roller-system for the undulating movement [1900]

horizon. Oran is reached at sunset. The sea presents an ever-changing picture; now we encounter some fishing-boats, then come large passenger steamers, and finally there emerges an imposing squadron of warships, including ironclads, battle-cruisers and torpedo-boats;

The Trans-Siberian Railway panorama: the four screens, the railway carriages and the spectators [1900]

completed—are shewn.

During a real railway journey every traveller has noticed that the nearer the scenery is to the train, the faster it seems to pass before one's eyes. Accordingly, a horizontal transmission belt to which sand and boulders have been attached is situated in close proximity to the carriages, and is driven along at a speed of 1,000 feet per minute by an electric motor. Behind it, a low, vertical screen on which shrubs and brushwood have been painted travels past at 400 feet per minute. The next slightly higher screen, showing the more distant scenery, passes before the spectator's eyes at 130 feet per minute. Finally, the remote background consisting of mountains, forests, clouds and towns and villages, painted on a screen 25 feet high and 350 feet in length, crawls along at 16 feet per minute. As the whole building is merely 200 feet long and the screen cannot be rolled up—it must always move in one direction only—it passes as an endless belt in reverse direction at the back of the block.

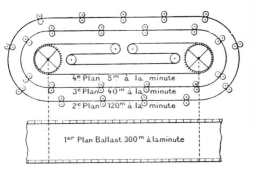

On the screens, the French decorator Jambon has depicted the capital cities along the route such as Moscow, Omsk, Irkutsk, the shores of Lake Baikal, the Great Wall of China, and finally China's capital, Peking. The whole journey takes forty-five minutes. A noteworthy feature is that the scenery will never repeat itself exactly because, due to the difference in speed between the screens, the overall picture presents an infinitely variable combination of scenes.

THE MAREORAMA

When the screen of the above-mentioned panoramas unrolls and the spectator experiences the sensation of travelling along he nevertheless feels no bodily movement, and in this respect the illusion is far from perfect. The Mareorama eliminates that drawback in that the spectator himself is in motion and actually feels the roll and pitch of a ship while making a sea voyage by way of Nice, the Riviera, Sousses, Naples, Cape Pausilippe and Venice to Constantinople.

The plan for the Mareorama presented two problems: two screens, each 2,500 feet long and 40 feet in height, were to be unrolled, and a double, swinging movement was to be imparted to the spectator's platform which was shaped like a ship. The mechanism required for this was conceived by Mr Hugo d'Alesi, a well-known painter who specialises

The Mareorama: the mechanism for the unrolling and rolling up of the screens [1900]

in rendering the most beautiful vistas for the posters of the large railway and shipping companies. The screen is also his work, painted after the sketches made by him during a voyage of one year especially made for the purpose. For eight months, a team of painters worked under him to transfer these to the 215,000 square feet of screen which was to be unrolled before the visitor's eyes.

One of the screens moves on the port side, the other on starboard. Both are coiled upon cylindrical reels situated near the ends of the building where they are concealed from the view of the ship's passenger by sails and ornaments. In the engraving one of the screens has been removed in its entirety to show the mechanism for the movement of the two screens, as well as one of the vertical cylinders round which the second screen will be rolled. These extremely heavy cylinders are supported by floats in a water-basin. To simulate the roll and pitch of a ship, and to impart

125

The Mareorama: the mechanism for the rolling and pitching [1900]

these movements to the boat-deck carrying the spectators, it is supported by a system of Cardanic rings similar to that used for accommodating ships' compasses. This involves the use of floats in water, hydraulic piston engines, and pumps driven by electric motors.

Few visitors to the Exhibition will be able to resist the temptation of this opportunity to make an inexpensive voyage which involves no hazard whatsoever, yet is so natural that one can even make acquaintance with the less agreeable sensations to which passengers on board ship are likely to be subjected. While this may also deter many, it is a reassuring thought that even on the high seas, amid the raging elements, one can get out and tread on terra firma at any moment.

CINEORAMA AIR-BALLOON PANORAMA

When watching the aforementioned panoramas, the spectator was imagined to be in a railway carriage or on board ship; but a third means of conveyance, the balloon, is available to provide a new panorama such as presents itself to the air traveller when taking off, hovering and alighting in his craft.

Rare indeed are those among us who have ventured on a flight in a balloon and who have experienced the indescribable sensations which must assail the airborne traveller as he rises aloft and floats along in the immense ocean of the sky, far above the earth teeming with all sorts of creatures, finally returning to terra firma at the end of the journey.

If it were possible to display a panorama which would convey the sensation of air travel without entailing any of the real dangers, it might justly be expected that the public would flock in crowds to see and enjoy such an unusual spectacle. Prior to the Paris Exhibition this was not possible, although the simultaneous projection of a number of slides on to a circular screen surrounding the spectators had already proved successful. However, this was merely panoramic projection and not a cinematographic panorama; the 'living' pictures of the cinematograph had still not become a practical reality. In the 'all-around' projection of at least six slides in as many lanterns, the trouble always was how to prevent the spectators from seeing the joins between the various pictures. This was achieved only by accurately setting the diaphragms and small screens of variable and adjustable size. How much more difficult in the case of moving pictures!

Monsieur Grimoin-Sanson, a French engineer, appears to have found a solution. For his exposures he uses a scaffolding 9 feet in height carrying ten cinematographic cameras, each covering an angle of just over 36 degrees to ensure that the projected pictures overlap slightly. The cameras are operated simultaneously by turning a crank which is connected to each by a system of gears. Having recorded a number of interesting panoramas on terra firma in various countries with this device, the operator did not hesitate to place it in the pannier of a balloon despite its weight which exceeded 1,000 pounds. Early in May 1890, the balloon rose into a slightly overcast sky from the Tuileries in Paris. Immediately upon the cry of 'Let her go,' the entire machine was put into operation and continued until the balloon had reached an altitude of 1,500 feet.

As they see the earth recede, the spectators

of these cinematographic pictures experience the sensation of rising into the air. By reversing the direction in which the successive frames are projected, the illusion is created that the balloon is descending and landing.

A polygonal-shaped hall with a diameter of 100 feet has been built for the projection of the film strips, its walls consisting of ten screens 31 feet square. A round bowl of reinforced concrete stands in the centre and on top of this, there is a platform on which

The Cineorama air-balloon panorama: total view of the interior [1900]

the spectators stand. Above their heads, a huge sail is fixed to the ceiling of the hall by means of nets: this is the 'balloon' which is supposed to carry the visitors aloft.

Along its circumference the bowl has ten apertures through which the objectives of the ten projectors protrude. An electric arc-lamp of 10 ampères is placed at the back of each projector. To shield the operators from excessive heat, each lamp is housed in a tin box fitted with a funnel to dissipate the warm air generated by the lamp. In addition there is a fan which blows cold air continuously into the bowl.

The strips of film containing the various cinematographic exposures have been glued together to form a single ribbon over 1,300 feet in length, thus permitting continuous projection for more than six minutes. Each frame on the strip is approximately 2 inches square and is projected on to a screen of 915 square feet, resulting in a linear magnification of more than 180 and an area magnification of about 34,000. For a show of more than six minutes, ten strips each 1,300 feet in length are used, producing a total of well over 80,000 images!

It is reported that so far more than 50 million people have visited the Paris Exhibition. It is not known how many of them have undergone the sensation of a 'balloon ascent' in the Cineorama.

———

TELEPHONE

Bell's telephone [1877]

GRAHAM BELL'S TELEPHONE

1876 has been a memorable year in the history of inventions. To commemorate the centenary of the foundation of the republic of the United States of North America, a World Exhibition was held at Philadelphia. As is customary at such exhibitions, instruments and machinery of various kinds could be admired there, bearing witness to the creative spirit of America's engineers and scientists and the imaginative flair of her most skilful tradesmen. Among all these inventions, there was one which not only contributed most to the fame of the Exhibition, but established the name of the United States as a nation of brilliant inventors. Yet it was merely a simple device which its

Professor Graham Bell delivering his speech [1877]

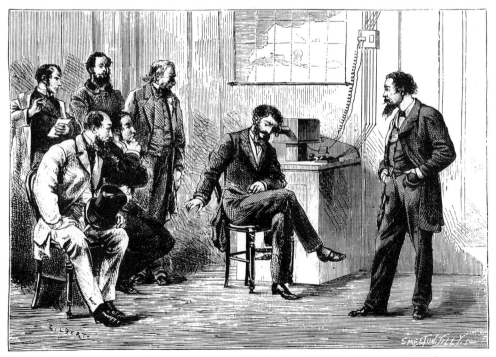

Boston audience listening to a speech by Graham Bell 15 miles away [1877]

inventor, Alexander Graham Bell, publicly presented there for the first time under the name of 'telephone.'

At first, the instrument attracted only slight attention due to its ungainly appearance and seeming lack of importance; and indeed little else could be expected, considering the poor show that Bell's telephone must have made, in the company of such a multitude of steam-machines and other complex pieces of machinery. All this changed, however, when the instrument drew the attention of no less a personage than Don Pedro, the Emperor of Brazil. And when it became known that the telephone could speak, almost as perfectly as the human mouth, rendering the spoken words audibly even at a considerable distance, its fame spread like wildfire. From then on the telephone was an object of general admiration.

In the Old World, where the scientists believed that they knew all there was to know about the theory of sound, many fruitless attempts had already been made to resolve the same problem. Many therefore held the view that no solution could ever be found, and people were naturally in doubt as to the accuracy of the reports about Bell's

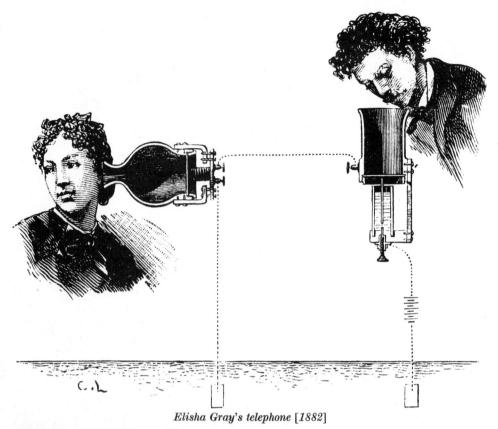

telephone that reached Europe from America by telegraph. Not until W. Thomson, the well-known English physicist, submitted his report to the British Society for the Promotion of the Sciences was the information—hitherto regarded as mere exaggeration—accepted.

During a meeting held by the Society at Glasgow in September 1876, Thomson said:

'At the Philadelphia Exhibition, I heard the words "To be or not to be—there's the rub" spoken through a telegraph-wire, and the electrical pronunciation emphasised the ludicrous nature of these monosyllables, making them sound even more laughable; the apparatus also recited excerpts taken from newspapers from New York. My ears perceived all this very clearly; the words were pronounced very distinctly by the thin, circular plate forming the armature of an electromagnet. It was my colleague from the jury, Professor Watson, who spoke the words in a loud voice at the other end of the line; for that purpose he brought his mouth up to a tensed membrane fitted with a small piece of soft iron, which performed movements corresponding to the vibrations of sound in the air, in front of an electromagnet in the electrical circuit.

'For this discovery, the wonder of wonders

Elisha Gray's telephone [1882]

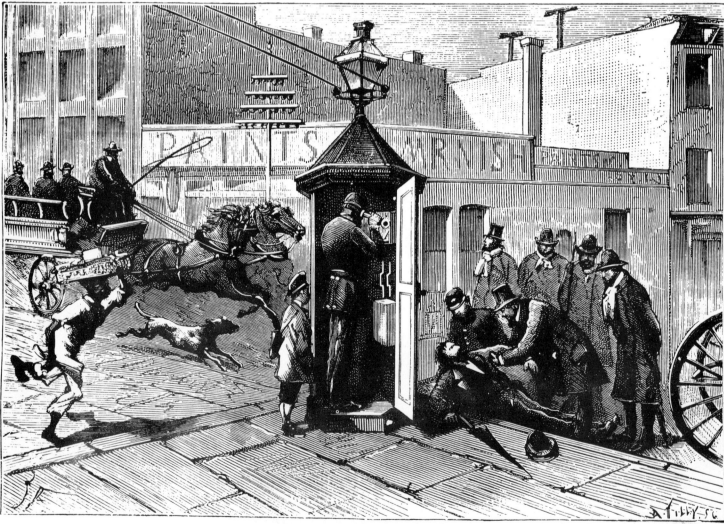

The Chicago police has called up a police vehicle by telephone for a man hurt in an accident, details of which are passed on to the police-station [1878]

130

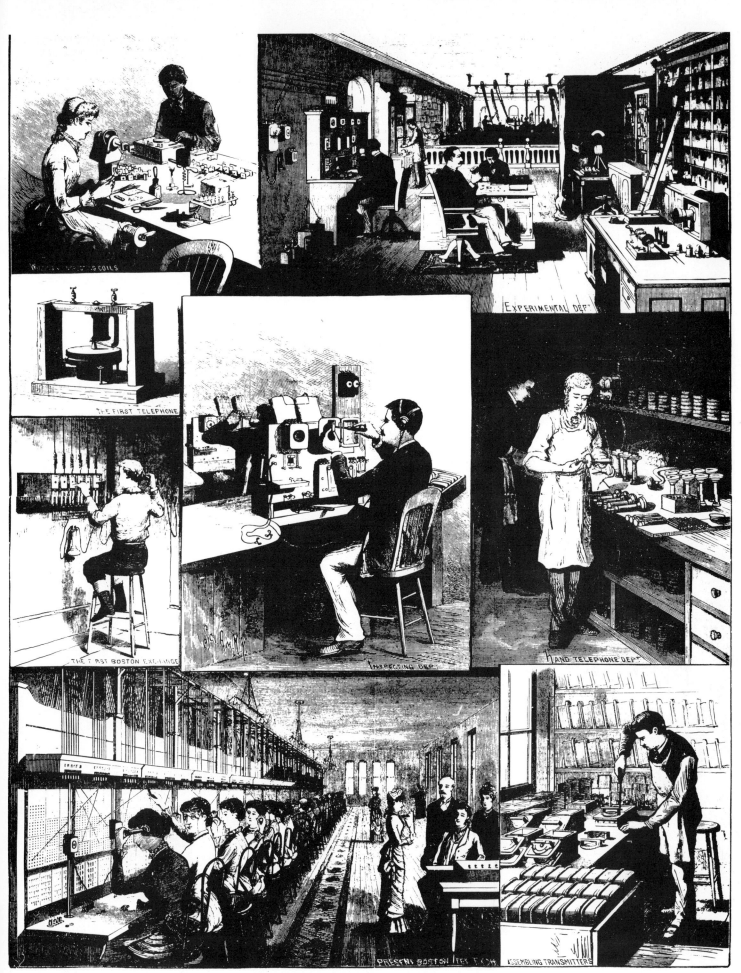

The American Bell Telephone Company [1884]

The telephone exchange at the Avenue de l'Opéra in Paris [1882]

practical shape which Bell gave it six months later after making countless tests. This new apparatus no longer requires an electrical battery, and consists of a powerful horseshoe-type magnet, the arms of which are provided with wire coils. This magnet is placed before a thin, soft-iron plate. The apparatus was presented at the Essex Institute at Salem, Massachusetts on 12th February 1877; a discussion held at Boston, 14 miles away, before an identical apparatus was made audible at Salem to an audience of 600 persons. Although the words spoken could be understood only by those who were in close proximity to the machine, the members of the audience present in the hall at Salem could still distinguish the voice of the speaker in Boston. Small wonder that, after such results, the attention of the whole world was focused on Bell's telephone.

In the meantime, much larger distances have been bridged by the new device. Recently, Bell in New York spoke to Professor Watson in Boston, 250 miles away; they had chosen a Sunday for this experiment in order to avoid being disturbed by telegraph operations on neighbouring lines. It is also possible to telephone messages by undersea cables. The longest distance so far spanned by undersea cable is between Dartmouth and the Isle of Guernsey, which are 60 miles apart. The velocity at which the sounds of the telephone are transmitted is, so to speak, infinitely great.

of the electric telephone, we are indebted to one of our young countrymen, Mr Graham Bell from Edinburgh, now a naturalised citizen of the United States. One cannot but admire the boldness of the invention which

has, by such simple means, solved the complex problem of reproducing the composite sounds of voice and speech by electricity.'

The remarkable device which earned Thomson's admiration still lacked the

The telephone exchange in New York [1880]

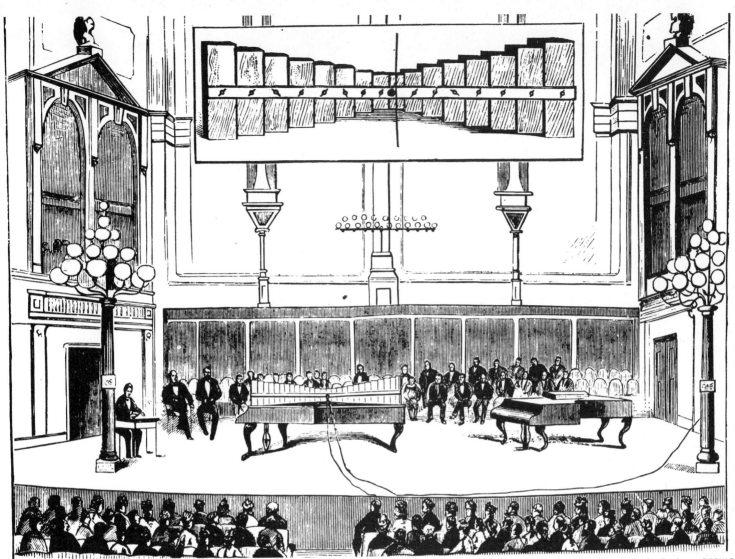

A demonstration of Gray's musical telephone [1877]

MUSICAL TELEPHONE

New York recently witnessed an impressive demonstration given by Professor Gray. It concerned his invention, the musical telephone, which is built upon the framework of a grand piano. Comprising two octaves, the instrument consists of sixteen iron tongues of varying length and an equal number of electro-magnets and wooden jackets. The combination of tongues and electro-magnets has the effect of a tuning-fork, the vibrations of which are reinforced and audibly reproduced by the wooden sound-box. The vibrating metal tongue causes electrical impulses to be generated in a wire coil connected to a telephone wire. In this way, these impulses actuate a similar instrument at the other end of the line a long distance away. Recently, a melody played in New York was transmitted by this system to Philadelphia where it was heard by a large audience. The sound of the instrument is loud and pleasant.

RENDERING OF MUSIC BY TELEPHONE

McDermott's telephone [1880]

Some time ago we reported on the theatrical and brass-band performances diffused to a large audience by telephone. A slightly modified system is now used to convey the sounds produced by a quartet of 'Mirliton'-players to the visitors to the Telephone Pavilion near the Eiffel Tower, Paris.

The sound transmission system consists of two basic elements: the transmitter and the receiver. The transmitter has a mouthpiece into which the flautist pipes the sounds of his instrument, causing periodic interruptions of the current passing through the device, which is included in the electric battery circuit of five receivers; the twenty receivers mounted on a panel in the Telephone Pavilion thus form in reality four independent sound-transmission groups, the vibrations of each of them being absolutely dependent on the performance of the individual artist.

At the time of writing, it is difficult for us to make any definite statement on the musical and artistic value of these telephonic transmissions of music but, remembering the success scored by this system on its début at the World Exhibition of Electricity held in 1881, it is only to be hoped that it will pave the way for carrying the human voice over long distances with the same intensity as the present group of 'Mirlitons'. Although the results so far obtained are not brilliant, we must not desist from our efforts. After all, only fifteen years ago the telephone did not exist whereas now, it spans the entire world. . . .

The rendering of music by telephone: the receivers [*1889*]

THE THEATROPHONE BRINGS MUSIC INTO THE HOME

Only a short time after the telephone had commenced its triumphant progress round the world, attempts were already being made to use the new instrument for transmitting music over great distances. Even then men dreamed of a time when, without leaving the comfort of their home or study, it would be possible to enjoy a Beethoven concert which was being performed elsewhere, most likely several miles away. These illusions have gone less than half-way to being realised. It is true that, mainly at exhibitions, there has been an occasional opportunity to press a pair of telephones against both ears for a few minutes, and then one could perceive— generally in a most imperfect way—that a piece of music was being played at the end of the line, but nothing more than that. There was no question of artistic enjoyment in such transmissions of music over distances great or small. It was indeed amusing—a thing to wonder at, but there the matter ended.

If we may believe the *Electrical Engineer*, the Long-distance Company in America is now successfully transmitting quartets and quintets over distances of many miles. And not just so that only a few persons armed with ear-telephones can hear the music, but in such a way as to allow a reasonably large audience to listen to the performance with real pleasure. There are two main problems to contend with in the transmission of musical works by telephone. The first lies in ensuring that the sounds are picked up by the microphones provided for the purpose, and the second in reproducing these sounds loudly and clearly enough at the receiving station.

As far as the transmission is concerned, it has proved essential to use a separate microphone specially arranged for every instrument or for each voice, for otherwise the result at the receiving station will be a mixture of sounds in which the relationships in the intensity of the various notes produced at the transmitting station will either be reproduced very badly or practically not at all. Another great drawback is that the sound produced by the telephones is generally so faint that it becomes necessary to oblige each listener to cover both ears with telephones during the performance. If concerts by telephone are eventually to achieve any measure of success these cumbersome conditions will have to be eliminated.

In Figs. 1 and 2, p. 137, we can see the procedure adopted by the Long-distance Company in America. Fig. 1 shows five musicians performing a quintet. They have taken up position before four separate microphones fitted with special cups which serve to concentrate the vibrations of the air. The cornet and the double-bass share one microphone between them. The four microphones are now placed 'side by side' in four separate circuits fed by a battery of accumulators.

In addition to the microphone, each of these circuits also comprises the inductive or primary wire of an induction-coil. The secondary wires of these inductive coils are then joined together one after the other in a circuit and the current thus combined is transmitted over the line to the receiving station. This is a current of variable intensity which faithfully unites the combined wave movement produced by each of the instruments so that the relative intensity of the notes played by each of the instruments used is heard to perfect advantage.

Figure 2 shows a part of the concert hall where the auditorium is located. As can be seen, it is not the normal Bell telephone which is used here but the loudspeaker telephone invented by Mr Edison. This enables a large number of people to hear the music at one time. According to the *Electrical Engineer*, by placing six loudspeaker telephones one behind the other and distributing them over the walls and chandeliers in the hall, this method has been successfully employed to let more than 1,000 people listen simultaneously to a concert being performed 250 miles away.

A company has recently come into existence in Paris which provides facilities for listening by telephone to vocal or instrumental performances given in one of the city's many theatres. Thus in many places in France, and even in other countries which are linked to Paris by telephone—including Brussels and London—it is possible to hear the performance from a Parisian theatre by this means. Figure 3 shows the instruments with which people can listen to the chosen performance at different public places. Various of these are to be found in the salons of the great Parisian hotels. One need only

The rendering of music by telephone [1889]

slip a one-franc piece into an opening in the front of the 'theatrophone' to be able to listen for ten minutes at a time. Subscribers to the telephone service do not require these devices as they can listen with their normal telephone.

The central exchange (Fig. 4) is situated in the Rue Louis-le-Grand and it is here that all the conducting wires come out. Every evening a young girl sits at her post and makes the connections as required. There are three separate lines. First, there are the lines which connect the microphones in the theatres to the central exchange; then come the lines which connect this exchange to that of the State telephone service, from where telephonic communication can be established with all subscribers in France or countries abroad; thirdly, there are the lines for the theatrophones at public places such as the salons of the hotels.

In the theatres the microphones stand on the stage beside the footlights and are fed by Leclanché cells. From these run lines which come out at the rosette (junction-box) and the commutator of the central exchange; all the lines which run to the State telephone service and to the places where the theatrophones are installed also terminate there. The lady operator need do no more than make the connections to the theatres requested by the subscribers. She does not have to wait for requests for the automatic theatrophones for they are ready to start functioning as soon as the theatres are open.

The listener must know whether the instrument is ready and to what theatre he is connected. For this purpose there is a plate fitted with an indicator which shows all the theatres where performances are taking place and which also bears the word 'entr'acte'. Another plate with the same information, but somewhat larger in size, hangs at conspicuous places in the hotels, cafés or restaurants.

The indicators on these plates are controlled by an electromagnet which receives current by means of a generator at the disposal of the operator in the central exchange. This can be seen in Fig. 4, p. 135. It is shaped like a little wheel with a crank which is rotated in order to open or close the circuit and this also happens in the case of the indicator telegraph. This telegraph is served by a separate line, the current for which is

The central exchange of the theatrophone in Paris: fig. 4 [1893]

derived from the lead for the electric lighting system.

On her arrival at the exchange at about eight o'clock in the evening, the operator first finds out which theatres are open. She then selects one and connects it to all the lines of the public theatrophones. This done, she works her telegraph to set the indicator to the name of the theatre she has chosen. As soon as she notices that there is an 'entr'acte,' or interval, she connects the theatrophones with another theatre and changes the name in all indicator-plates by turning the crank.

It is obvious that all the theatrophones on the same line will also receive the programme from the same theatre; meanwhile, it is also possible to make each line independent of the other by means of a commutator so that, for example, one can receive the Opéra-Comique while the other receives the Opéra.

As can be seen from Fig. 3, there are some fifteen small indicator-plates above the commutator in the central exchange. Each of these is connected to one of the lines mentioned and the indicators operate in parallel with those of the receivers. Thus the lady operator can see at all times what she has telegraphed and can change the indications at the appropriate moment.

So the theatrophone is here. What is still lacking is the 'telephote' which would also enable us to see the actors or singers at all the distances over which they can now be heard. An ingenious device to locate the singing actor was invented and demonstrated by Ader as early as 1881: a microphone, situated to the left of the stage, was connected to a telephone intended for the left ear; another microphone placed to the right was linked with the telephone for the right ear by separate wiring. To the listener, the effect to each of his ears was different in much the same way as the effect of the stereoscope for the eyes. [Stereophonic!]

Theatrophones in the lobby of a hotel in Paris: fig. 3 [1893]

136

The transmission station in New York: fig. 1 [1893]

The audience listening to Edison's loudspeaking telephone: fig. 2 [1893]

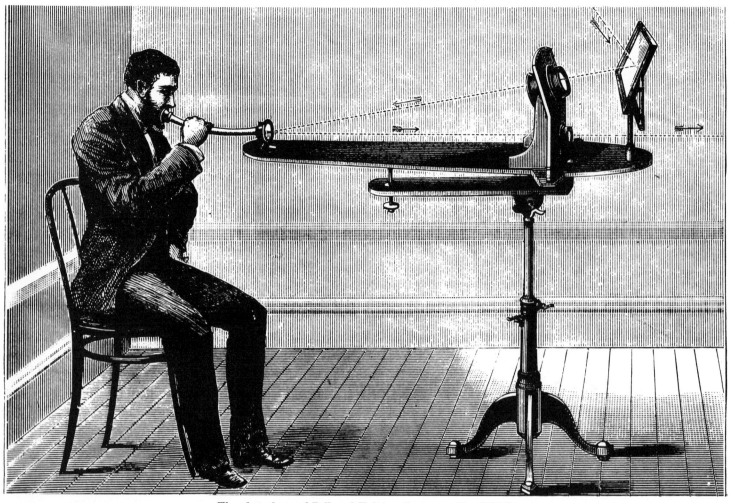

The photophone of Bell and Tainter: the transmitter [1881]

PHOTOPHONE, OR SPEECH BY LIGHT

One of the most interesting inventions of recent months is the photophone or 'light speaker.' On 26th April 1880 Alexander Graham Bell, the famous inventor of the now familiar telephone, announced that while working in conjunction with Summer Tainter he had made some discoveries which had finally led him to manufacture an instrument with which sound can be transmitted and propagated by means of light.

The photophone consists of a transmitter and a receiver. In the case of the receiver, the rays of a powerful light source, such as sunlight or electric light, are reflected by a flat mirror into a system of lenses which forms the light into a beam and throws it on to a silver-coated glass plate. When someone speaks into the mouthpiece of a rubber tube the silver-coated plate picks up vibrations which cause the reflecting surface of the plate to change shape. This in turn affects the strength of the light-rays which are concentrated into a parallel beam by a second system of lenses and then reflected towards the receiver.

This latter instrument, situated some distance away, consists of a hollow, parabolically ground mirror of silver-coated copper. A cylinder of copper plates separated by strips of mica and filled with the metal

selenium is placed in the centre of this mirror. The resistance of this device changes in accordance with the quantity of light which strikes it. The selenium between the edges of the copper plates is incorporated in a circuit consisting of a nine-cell battery and a double telephone which reproduces everything the speaker pleases to say into the rubber tube.

Bell and Tainter have devised and tested some fifty different designs for the photophone. The greatest distance over which they succeeded in transmitting the spoken word was 213 metres. Tainter stood on the tower

of the Franklin School in Washington with his transmitter and Bell remained at the window of his laboratory with his receiver. When Bell put the telephone to his ear he heard Tainter say quite clearly: "Mr Bell, if you can hear what I am saying come to the window and wave your hat."

The photophone can also transmit songs with great purity of tone. By introducing a perforated disc into the beam in the transmitter, musical notes are obtained in the telephone of the receiver. If there are thirty holes in the disc, and it rotates ten times per second, a pitch of 300 vibrations per second is set up. The possibilities which this holds for the world of music are undoubtedly remarkable.

TELEGRAPHIC COMMUNICATION BETWEEN MOVING TRAINS AND STATIONS

The amazing speed at which the multitude of trains fly past each other or follow one another nowadays makes the dangers of a railway journey much greater than they were when people were content with travel at much lower speeds and with fewer travelling facilities. Fortunately, however, human ingenuity has managed to keep pace successfully with these developments and safety

devices have increased in step with the growing dangers, so that the number of railway accidents is relatively decreasing rather than rising.

The latest of these devices, which has already been tried out in America and is reported to be completely reliable, consists of a telegraphic or telephonic link-up between moving trains and intermediate stations,

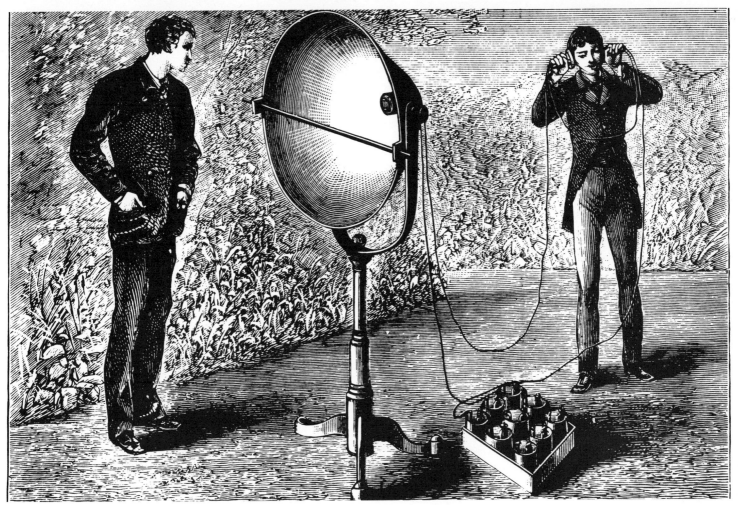

The photophone of Bell and Tainter: the receiver [1881]

The Edison system of railway telegraphy [1886]

and even with other trains rushing past.

The importance of such an invention will readily be recognised. The station-master can find out at any moment the position of the train he is expecting. He can inform the driver of possible danger and stop the train at any given point. On the other hand, the guard can notify the station of any accident which occurs so that the appropriate measures can be taken there. How many collisions might have been avoided if only this method of communication had been known of sooner.

The system of Lucius J. Phelps has already been in use between New York and New Haven for one year. It was recently the means of preventing an accident for it informed the guard of a moving train that a train ahead had broken its couplings and that runaway coaches were making the line unsafe. Phelps's system is based on induction currents which are set up in a conductor when electric currents flow and are interrupted in an adjacent, parallel conductor. For this purpose a well-insulated conductor in a wooden cylinder is placed between both rails and connected to earth, on the one hand, and to the signalling equipment in the station, on the other. A coil of copper wire is wrapped round the telegraph coach in ninety well-isolated windings with a total length of 7,800 feet. The bottom windings, 3,250 feet long, run parallel to the insulated conductor between the rails. A battery with four to six cells is incorporated in the electrical circuit of this coil. A special signalling key enables the operator to send currents in two opposing directions through the wire coiled round the coach. When these currents are generated, disappear or are reversed, induction currents are set up in the conductor between the rails which can be received at the station. Conversely, current surges can be sent from the station along the conductors between the rails and these create induction currents in the coil round the telegraph coach. These then activate a sensitive relay termed a 'sounder'—a device which produces clearly audible signals. Even when the train attains its highest speed the ticking of this apparatus can still be heard at a distance of 10 feet.

The latest system of railway telegraphy is that developed by Mr Edison and Mr Gilliland. It too is based on the principle of induction, employing a telegraph-line along the track which acts as one plate of an elongated condenser. The other plate of this condenser is formed by a 30-centimetre-wide strip of copper fitted along the train on insulating ebonite plates. When the coaches are coupled, these strips are interconnected by means of flexible metal conductors. Due to the relatively low capacitance of this unusual condenser, high voltages are required, and these are obtained by using devices known as 'transformers.'

The telegrams are received in the form of Morse signals which can be heard in two telephones kept continuously at the ears of the official in the train who is still free to use his hands—see the illustration p. 139.

The usefulness of such communication is so self-evident that we do not doubt that it will be introduced on all trains in due course.

Telegraph and telephone wires in Philadelphia [1890]

Marconi with his transmitter and receiver [1898]

TELEGRAPHY WITHOUT WIRES

A lecture on wireless telegraphy was recently given in Toynbee Hall, London, by Mr W. Preece, Head of the Electricity Department of the General Post Office. He mentioned that when the cable between the mainland and the Isle of Mull was interrupted, he himself had succeeded in transmitting telegrams by purely inductive means over two telegraph-lines situated on the coasts 4½ miles apart.

According to Mr Preece, however, an entirely new system of wireless telegraphy—superior to his own—has been invented by a young Italian, Signor Guglielmo Marconi. Though the fact that the Italian postal authorities would not take up his ideas may appear slightly suspicious, the British Post Office are now seriously studying them, and will spare no expense to reach a decision in this matter. Mr Preece said that he personally placed the highest confidence in these efforts in which use is made of Hertzian waves of up to 250 million per second.

During his lecture, a few experiments were made with Signor Marconi's apparatus in the hall; its mode of operation was not, however, shewn, secrecy requiring to be observed in connection with the patent rights. The transmitter was housed in a cabinet at one end of the hall; an alarm-clock was at the other end, and there was no wire connection between the two. When the transmitter was operated, the bell sounded! If the invention were not endorsed by an acknowledged expert such as Mr Preece, one would be inclined to think there was some trickery involved.

From information given us by someone who claims to have spoken to Marconi himself we note the following details: born at Bologna, Mr Marconi is now twenty-two years old. His father is an Italian gentleman of means, his mother is an Englishwoman of good family. Since his youth he had been dedicated to electrical experiments. Having occupied himself for some time with Hertzian equipment, he discovered in September 1896 that he was able to transmit Morse signals with the Hertzian waves over a distance of ¾ mile. Tests performed in the Head Post Office in London revealed that these Hertzian waves could penetrate through seven brick walls. From this building, Marconi managed to transmit signals over a distance of 4½ miles. That distance could be much greater because it is entirely dependent on the amount of energy available and on the size of the equipment. Mr Marconi believed that he would eventually succeed in signalling from London to New York. The energy required for this was estimated by him to be between 50 and 60 horsepower, the cost of erecting the two stations being less than £10,000.

At the time of writing, Messrs Marconi and Preece are engaged in experiments at Penarth to establish communication between lightships and islands with lighthouses in the Bristol Channel. Mr Marconi claims that ships may be equipped with apparatus capable of indicating the proximity of other ships as well as the direction in which they are moving, both by night and in foggy weather. He has already succeeded in igniting gunpowder from a distance of 1½ miles. Marconi believes that with the aid of Hertzian waves, the magazines of warships may be ignited by signals transmitted from lighthouses, and that one ship may blow up another by similar means in future wars. All these claims and contentions have created high expectations, and it is hoped that we shall soon hear more about these promising developments.

EXPERIMENTS ACROSS THE CHANNEL

In the highly successful experiments carried out in 1898 near the Isle of Wight and the Irish coast, Marconi even managed to send a report on a sailing regatta from a steamer by

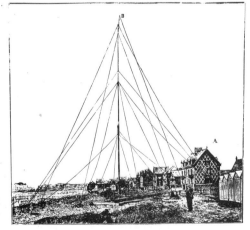

200-foot high reception pole [1899]

Marconi's receiver and transmitter during wireless telegraphy across the Channel, Marconi reading the dots and dashes written by a Morse recorder connected to the receiving equipment [1899]

ether telegraphy to various newspapers. The inventor is now conducting new experiments with wireless telegraphy between England and France. The pictures show the 200-foot-high reception pole or air conductor B on the French coast near Wimereux (Boulogne) as well as the receiver situated at A. During our visit to the site of the experiments on the French coast, we were intrigued when a signal came back from across the Channel almost immediately after the operator there had been called up at our request. We were most surprised to see the return telegram transmitted by the electrical waves—sucked up, as it were, from the atmosphere by the reception pole—clearly appear in Morse code on the emerging paper strip. The record for the longest distance so far spanned by wireless telegraphy, is now 42 miles. This impressive success was achieved by Marconi during tests conducted between the British coast and a French vessel.

The latest news is that, by 'tuning' the transmitting apparatus to specific wavelengths, Marconi has managed to eliminate mutual interference between transmitters operating at the same time. Only by means of this 'tuning' will wireless telegraphy become possible on a large scale.

SLABY WITNESSED THE EXPERIMENTS

Just as we were going to press, we received a report by Professor Slaby, a German scholar who was present at the experiments on the

Bristol Channel. Professor Slaby relates in lively fashion, how Messrs Preece, Marconi and himself huddled together, seeking shelter from the violent gusts of wind in the wake of a large wooden box, in keen expectation of what would happen. And how, following the signal given with a flag, the Morse apparatus was heard ticking away, clearly but invisibly transmitting messages from the coast, the outline of which could barely be seen.

DUSSAUD'S TELEOSCOPE

The fact that the electrical conductivity of selenium varies with the amount of light it receives has, in recent years, given a new impetus to the efforts of the engineers working on a method to see at a long distance by means of a simple electric wire.

Monsieur Dussaud, inventor of the microphonograph, the sound-amplifying device which is so helpful to the deaf, has now constructed an apparatus for seeing at a distance, as shewn in the accompanying drawings.

Fig. 1, p. 142, shows the person whose image and movements are to be reproduced, standing in front of the apparatus called 'transmitter'. It consists of a camera obscura B, at the back of which there is a revolving shutter C. Its disc is provided with small perforations forming a helix. A screen carrying strips of selenium is placed directly behind the shutter, the selenium forming part of an electric

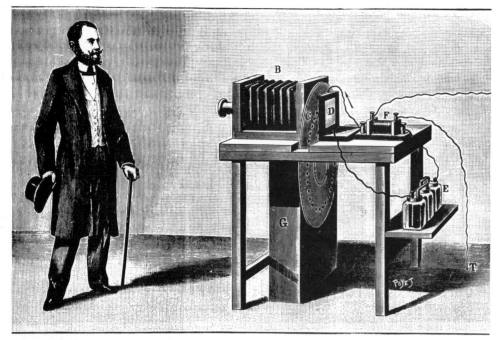

Dussaud's teleoscope: fig. 1, the transmitter [1898]

move in relation to a second, identical but fixed plate L which is placed behind it. Buffer glasses protect the plates against vibration caused by outside sources.

An arc-lamp produces a powerful beam of parallel rays of light, M. Directed at the two plates, the beam's intensity is reduced—throughout its cross-section—in direct proportion to the amount of current passing through the secondary windings of the coil. Since, however, the beam is forced to pass through a shutter O which is identical and rotating synchronously with the shutter C of the transmitter, this intensity of light is projected on the screen by an optical system P at a point which exactly corresponds with that of the image in the transmitter which has the same intensity. Since the shutters C and O perform a complete revolution in $\frac{1}{10}$ second, all the parts of the image successively act upon the selenium during that brief lapse of time, producing in the electrical circuits fluctuations which cause the intensity of the light-beam to vary accordingly. Passing through shutter O, this beam produces more or less bright spots of light on the receiver's screen at the rate of 10 per second, which is adequate to give to the observer's eye the impression of a continuously moving picture.

Monsieur Dussaud is at present engaged in improving his system, and it is hoped that he will succeed in completing his experiments in time for the Exhibition to be held in Paris in 1900.

When, like the telephone today, the teleoscope will be part of our everyday life, we shall have the thrilling experience of being able not only to speak to our fellow men over a long distance, but to see them as well.

battery circuit which also includes the primary windings of an induction coil F.

Since the image of the person moving in front of the camera is produced on the screen in its back, the various parts of that picture emit rays of light which, having passed through the apertures of the revolving shutter C, strike the selenium strips D. The shutter is driven by a clockwork G of the type found in the Hughes telegraph.

As the strips of selenium are successively exposed to the rays of light—of varying luminosity—emanating from the picture, currents of varying intensity are generated in the primary windings of coil F. These currents produce corresponding currents but of higher voltage in the fine, secondary windings of the coil which are connected to earth and, through a conducting wire, to the receiver (Fig. 2) where they cause vibrations in the membrane of a very sensitive type of telephone H. This membrane acts upon an opaque plate K which is provided with transparent grooves, causing it to

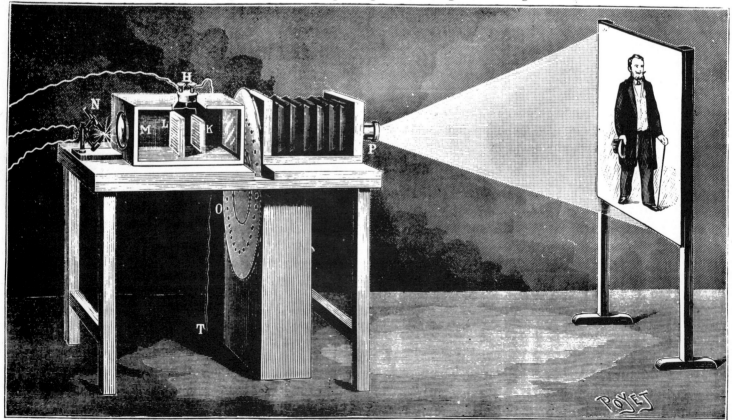

Dussaud's teleoscope: fig. 2, the receiver [1898]

MISCELLANEOUS

In tests carried out on 6th October 1883 at Arras, France, a war-mine exploded, killing and wounding several people

Interior spring mattress [1871]

AUTOMATIC FAN

Mr J. B. Williamson of Louisville, Kentucky, has invented a device which is capable of providing us with a pleasant and cool night's rest during the hot season. It consists of a clockwork mechanism driven by either a spring or weight to which a lever is attached, carrying long, narrow strips of suitable material. The engraving shows how the device, fitted above a sleeping couple, cools the air and drives away insects by the oscillating movement of the strips.

A NEW SHOWER-BATH

A Frenchman, Monsieur Gaston Bozérian, has constructed a shower-bath with which the bather himself can operate a pump by a simple walking movement, thus causing the water to circulate constantly so that only little water is required. This is of particular

A NOVELTY: INTERIOR SPRING MATTRESSES

We spend a third of our life in bed. It is therefore of prime importance to our health and our life-span to have a good bed at our disposal. Comfortable beds may be produced with birds' feathers and animal hair but these organic substances may also constitute a dangerous breeding-ground for bacteria, vermin and contamination.

To eliminate these drawbacks, a new type of mattress has been invented. Shaped like a flat and wide closed box, it contains several hundreds of helical springs, each of which is contained in a cylindrical envelope of some textile material. Sleeping comfort is ensured by a resilient layer of padding included in the top cover made of thick ticking. Rolled up for transport as shewn in our engraving, it forms a light, compact bundle of steel springs.

Automatic fan [1872]

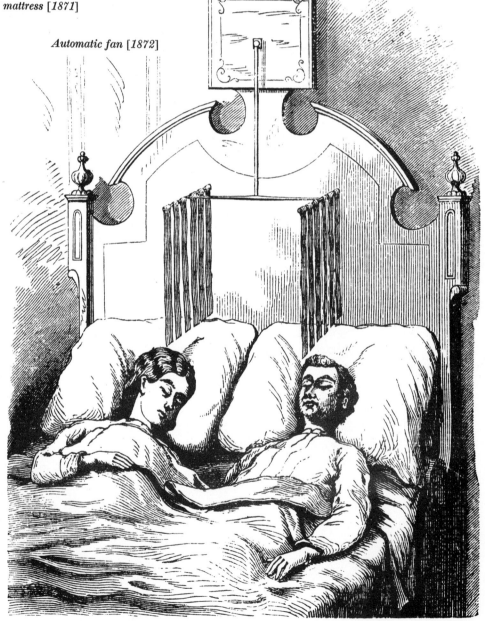

Apparatus to prevent snoring [1871]

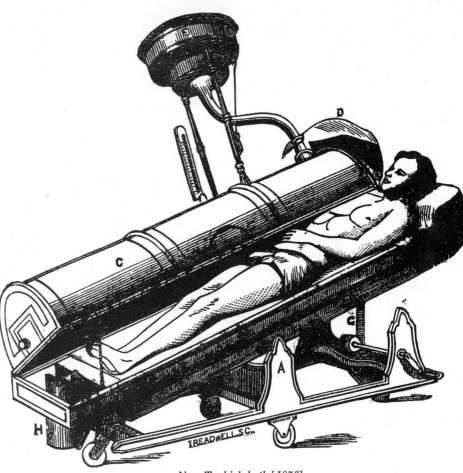

New Turkish bath [1873]

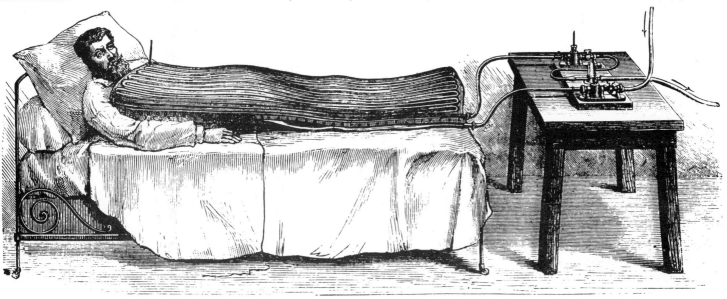

Bozérian's shower-bath [1878]

advantage if one wishes to take a hot shower-bath. Once the water has attained the desired temperature, a slight degree of heating is sufficient to maintain it at that level. A rail which can be held by the bather makes it easy to maintain one's balance when taking a shower-bath.

A NEW TURKISH BATH

The Turkish bath, as commonly practised, consists in placing the patient in an apartment heated by stove or pipes to a temperature of 100 to 120°F.; in a short time, as soon as the pores begin to open, the patient passes into a still hotter chamber, where there is a temperature of from 150 to 210°F. Here he remains until profuse perspiration is induced, and then, if he desires, enters a

room heated still higher. He then passes into a wash-room having a reduced temperature, is washed with warm water, then cooled with the spray-bath; he then plunges into a swimming-bath at the ordinary atmospheric temperature.

The Turkish bath is a beautiful luxury and has but one discomfort, to wit, the highly heated atmosphere of the perspiring-chambers. This is very oppressive to many persons; and to provide a portable bath as well as to overcome the difficulty is the

Water-cooled refrigerating blanket [1879]

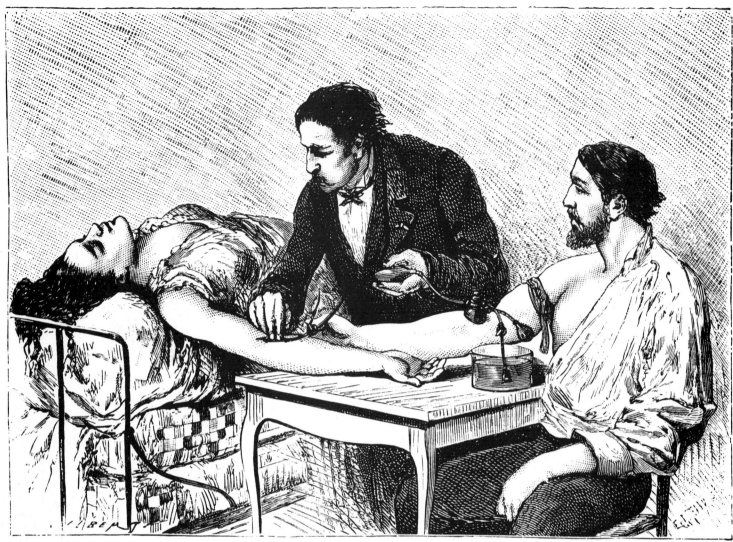

Direct transfusion of live blood, performed on 7th February 1882 by Dr Roussel

object of the present improvement, made public in the *British Medical Journal.*

A is the carriage upon which the bath rests, the wheels of which are so arranged that the whole apparatus can be turned completely round in a space little more than its own length. B, the frame and spring mattresses fitted with centres to the carriage, A, and forming the bottom of bath. C, enamelled metal cover, hinged to the frame B, forming chamber for heated air. D, waterproof and airtight apron to prevent escape of heated air at the top of the bath. E, cistern for shower bath. F, pillow with hinged headboard to turn up when the bath is not in use. G, rack and pinion for raising or lowering the bath to the level of a bed, for use of an invalid. H, heating apparatus.

The advantage over the ordinary public Turkish bath are these. The heat can be raised in less than ten minutes to 180 °F, and to the full temperature of 220 °F in fifteen minutes. The heat is obtained from gas, spirit or other suitable means. A shower-bath is attached, by means of which a copious discharge of tepid or cold water can be obtained, suddenly or gradually, at the pleasure of the bather or attendant. The head may, if required, be kept out of the bath in cool air.

Home-made inhaling-equipment [1894]

AEROTHERAPY

Judging from the multitude of novel remedies nowadays presented to the public, there must be an enormous increase in the ailments which beset the human family. There is an establishment on the River Rhine where the sick are healed by feeding them with grapes, the 'beer cure' is practised in Munich, while in Turin patients try to rid themselves of their diseases by drinking the still warm blood of freshly slaughtered animals.

In Milan, certain ailments are treated with . . . air. The invalid is taken to a cylindrical, comfortably furnished room into which chemically purified air is blown under steam-power and maintained at a pressure exceeding that of the open atmosphere.

Dr Carlo Forlianini, the discoverer of this therapy, claims that by increasing the pressure, the air is forced into even the finest branchings of the lungs, resulting in a much greater oxygenation of the blood and removal of specific obstructions in the bronchial tubes, while the respiratory muscles are also strengthened. Dr Forlianini also asserts that his method can be beneficial in the treatment of various blood and glandular diseases and pulmonary conditions.

The new therapy adds another blessing to the capabilities of the steam-engine: that of converting not only heat into pressure, but also pressure into health.

A HANDY PORTABLE SHOWER-BATH

The Frenchman, Monsieur Gaston Bozérian, has also constructed a collapsible shower-bath which can be stored so compactly in the round, flat tub that it can easily be taken on a journey. As the illustration shows, the water is pumped up by hand. A reservoir fitted right at the top permits the traveller to enjoy a brief, uninterrupted shower-bath after a short period of pumping. The perforated ring can be set at various heights and it can thus be adjusted to suit adults and children of differing stature.   [1880]

The Pneumatic Health Institute at Milan, Italy [1876]

Application of the McLeod system [1890]

HEATING, COOLING AND VENTILATING

The McLeod American Pneumatic Company, of New York and Boston, proposes to equip large buildings with a system of central heating which is not operated with hot water or steam but with air so that it will also be suitable for cooling during the hot season. For that purpose, a furnace to be constructed in the centre of the building will heat and circulate air through a series of pipes so arranged in a zigzag flue as to absorb all the heat from burning solid fuel or gas. Piped through the building, the hot air will be conducted through radiators transferring the heat to the ambient air; any excess of hot air may be allowed to escape whenever necessary or desired.

During the summer, cold water may be circulated through the furnace to feed cold air into the building. Special devices called 'thermostats' can be used for the automatic control of the temperature.

BREATHING EQUIPMENT

Every year, frequent accidents occur when workers employed in the distilleries have to descend into the fermentation cellars, and are overcome by obnoxious or intoxicating fumes in confined spaces, and the same to those who maintain and clean the sewers. Various devices have already been invented to provide rescue parties with pure air, thus enabling them to bring the victims safely into the open. The apparatus recently invented by Monsieur Vuaillet of Saint-Maurice in France, although very simple in design, appears to be successful. As shewn in the drawing, it consists of a mouthpiece having two spigots to which long, India-rubber tubes are connected: one for inhaling fresh

Breathing equipment [1900]

A NEW USE FOR THE INDUCTOR-BALANCE

The tragic event of President Lincoln's murder in Washington has revived the interest taken in the induction-balance invented by the British scientist, Professor Hughes. The device was in fact used in an attempt to locate the bullet in the injured President's body before it did its fatal work. It consists of two glass cylinders round each of which are wound two parallel coils of copper wire. One coil of each pair is included in a battery circuit in which there is a clock microphone. The other pair forms part of a closed circuit with a receiving telephone. The two glass cylinders, with their encircling pairs of coils, may be widely separated. The induction set up in the secondary or telephone circuit is balanced by the reversal of one of the secondary coils. The result is that the two induction currents neutralise each other so that, when the ear is applied to the receiving telephone, no sound is heard.

By placing ever so small a piece of metal in the electrical field of one of the glass cylinders, the balance is disturbed and the clock attached to the microphone is heard to tick loudly, thus indicating the presence of metal. The intensity of the sound increases as the pair of coils approaches the metal. This is by far the most sensitive method of tracing a bullet in a human body.

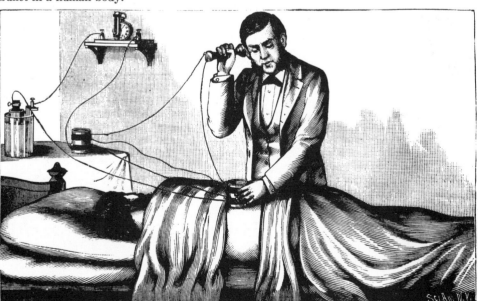

Hughes' inductor balance [1881]

Mechanical equipment for medical gymnastics [1897]

Feeding by means of a gastric tube inserted after excising part of the alimentary tract necessitated by blocking of the gullet [1888]

air, the other for exhaling. Valves incorporated in the spigots ensure that the user receives an adequate supply of fresh air and that the stale air is evacuated. A whistle is attached to the exhaling spigot to be used by the rescuer for signalling purposes. Secured by a rubber band passing over the head, the apparatus, being only 40 grammes in weight, does not impede the wearer's movements. Good results have been obtained with rubber tubes 100 feet in length. Nasal respiration is prevented by means of a clip placed on the nose.

A REMARKABLE SURGICAL OPERATION

Everyone knows how much progress surgery has made during the last few years and how many wonderful operations have been performed. Among these is the noteworthy surgical success recently achieved by a French doctor, Monsieur Terrillon, with a certain Mr X, a fifty-three-year-old man who had suffered for years from obstructions in the gullet. Using a rubber tube, Terrillon made a connection between the patient's stomach and abdominal wall, thus making it possible to introduce liquid nourishment such as beef tea and thin gruel directly into the stomach. Although Mr X is still troubled by pain—albeit to a somewhat lesser extent—the danger that he might starve to death has been averted by this operation and there is hope that his suffering will gradually decrease.

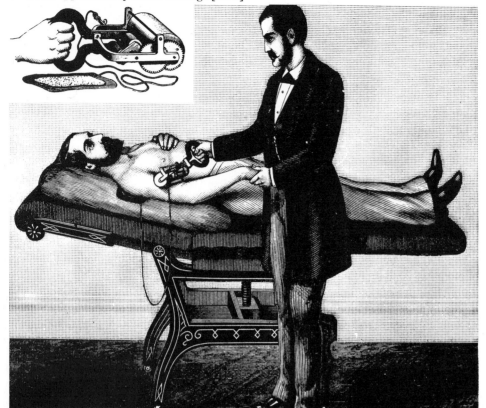

ELECTRO-MASSAGE

Dr John Butler of New York City has invented an instrument for electro-massage that can be operated even by unskilled persons. Energised by the motion of the leather-clad metal roller of the apparatus over the patient's skin, a built-in electro-magnetic generator revolving at the rate of 25 revolutions for 1 complete revolution of the roller, produces alternating currents of fairly high voltage. The poles of the generator are connected to blunt metal studs in the roller surface and to a suitable pad, likewise provided with a conducting surface, which is placed underneath the part of the body to be massaged. Combined with the stimuli engendered by the electrical impulses, the mechanical massage appears to be especially salubrious in cases of rheumatism, nervous exhaustion, neuralgia and paralysis.

CREMATION, OR THE BURNING OF CORPSES

When Christianity was introduced to the West the custom of burning corpses disappeared, either because the first Christians

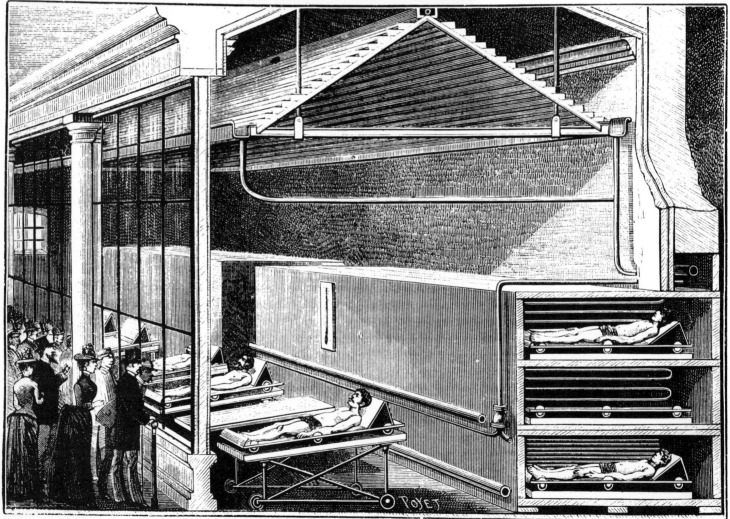

Every year, the Paris Morgue receives approximately 1,000 cadavers for identification. First the bodies are cooled down to –15°C and then stored at a temperature of –2°C [1887]

secretly buried their dead in order to avoid attracting attention or because the Apostles and Fathers of the Church wished to alter, at least in outward forms, everything which savoured of heathen customs. Thus arose the custom of burying the dead in the churches themselves, a custom which was preserved until a century ago. Or the corpses were buried in churchyards, in the immediate vicinity of the church, and so in the centre of the community. As a result, corpses piled up in the most densely populated parts of the towns and villages.

In these times when the so-called 'faecal' question is the order of the day, and when no costs are spared to remove rotting substances as swiftly as possible from the towns and render them harmless, it need hardly be said that rotting human corpses are also detrimental to health. This is why so many give the advice: 'Do not bury your dead, but burn them.' Those who support the burning of corpses, or cremation, consider the present method of burial extremely unhealthy, expensive and unaesthetic, a fact which even the greatest opponents of cremation must admit.

In most civilised countries there now exist associations for cremation which count among their members the most highly educated and enlightened men and which are vigorously and energetically engaged in spreading their ideas and winning acceptance for them. Nevertheless, they still encounter many obstacles, not only of a religious nature, but also in the field of law, since cases of poisoning would be more difficult to detect if only the ashes were left.

How is a corpse cremated? The first essential is that the body should be burned rapidly and completely, without any sound and above all without causing any smell. That is not such an easy matter when we reflect that a full-grown man weighs on average 154 pounds and that a body of this weight consists of 90 pounds of water, 14 pounds of ashy components and 50 pounds of organic combustible substances composed of about 19 pounds of nitrogenous substances and 31 pounds of fat. Very efficiently constructed ovens are required to burn all this silently and, above all, without creating any odour. Such an oven was not successfully constructed until 1876. At present, cremation ovens are in use in various towns and these satisfy all reasonable requirements. The oven shown here, in use in Berlin, is so situated that the entire cremation procedure can be observed through small peep-holes. The corpse is placed on a grid under which a container is placed to catch the falling ash. The oven is stoked from the front and the flames and combustion gases pass over the body, then travel downwards through a duct and pass through underneath the grid on which the body is lying, after which the combustion products escape through the chimney. By means of this arrangement the body is completely surrounded by flames and is rapidly and completely consumed. The oven is stoked with soft firewood to a heat of 600–700 °C (1112–1292 °F). It takes between one and a half and two hours to burn the corpse of a full-grown man and costs four or five shillings for fuel.

Meanwhile, the engineer Giuseppe Venini has constructed what he considers to be a better cremation oven in which cremation takes place solely by using gas flames mixed with hot air. As experiments have shown, this causes the body to burn more rapidly and completely than any other oven, converting it into absolutely odour-free gases.

The Milan cremation temple [1881]

A NOVEL METHOD
OF PRECIPITATING RAINFALL

Mr Daniel Ruggles of Fredericksburg in the United States of America has been granted the patent rights on a new method of making clouds discharge their contents, and hence precipitating rainfall, which we consider to be in no way chimerical. As many people have observed, an unexpected rain-shower often occurs after violent cannon-fire when the sky is heavily overcast. The discovery of Mr Ruggles is based upon this phenomenon. His object is to create shocks in the upper layers of the atmosphere which will bring about rain formation in the clouds. For this purpose he employs a balloon which carries aloft an explosive charge in the form of nitro-glycerine, dynamite, gunpowder or gun-cotton which is made to explode by an electrical discharge.   [1880]

COLLECTING AND
STORAGE OF ICE
IN AMERICA

The need for coolness and refreshment is one which has been felt in every age. Even in the sumptuous days of the Roman Empire, the rich did not shrink from spending enormous sums for the luxury of obtaining ice; during

POYET

the winters it was collected in Northern mountains, to be conveyed at staggering cost in straw-covered wagons or in galleys to Rome where it was stored in ice-cellars. In the barbarian Middle Ages there is no mention of such luxuries, but in recent times, with the progress of civilisation, there is again a growing demand for ice as a means to a more pleasant existence.

In everyday life, there is a steady increase in the number of businesses which make use of ice. Butchers and fishmongers, restaurants and hotels can not do without it. But in domestic life too there is a growing need for ice, especially for cool drinks on hot summer days. Ice is indispensable in factories where composite candles, nitroglycerine and paraffin-wax are manufactured, in distilleries, in dairies, in breweries and in many other businesses. An equally important application for ice is found in providing cold storage on board ships carrying fresh meat from Australia. In the practice of medicine, ice has become indispensable for local cooling and the time will surely come when cold air is generally conducted into hospital wards, theatres, and even into private houses in the same way as they are now heated by hot air in winter.

How is all that ice obtained? At first, efforts were made to satisfy the demand by cutting the ice from rivers, canals and trenches and storing it in ice-cellars. But in mild winters, this source of supply proved inadequate so that the ice had to be ordered from other, colder regions such as Norway.

In the United States of America, where the ice consumption in summer is amazingly high —in New York, as much as 1,450 pounds per capita per year—large quantities of ice are collected in winter time with little effort on the Hudson River and on the great lakes of Canada. About 100 miles from New York on

Sager's steam driven machine to saw ice into blocks of suitable dimensions [1883]

Wooden sheds in which the ice can be stored [1883]

The artificial skating-rink in Paris [1890]

the banks of the Hudson River, there are wooden sheds in which the ice can be stored until summer comes. Having double walls insulated by a layer of air, which is a poor conductor of heat, each of them provides storage for 50,000 to 100,000 tons of ice.

A few of these buildings are shewn in our picture and here you see the gathering of ice in full swing. The work commences as soon as the ice has attained a thickness of 7 inches. The places where sound, clear ice is found are marked by fir-tree branches and these areas are then divided into panels 3 feet square by a type of horse-drawn plough. The knives of the ploughs are pulled through the furrows several times so that these become deeper and deeper until the blocks thus formed are held together only by thin strips of ice. The blocks of ice are hewn loose with pick-axes and drawn to the sheds by horses. There, workmen place them, one by one, in boxes fixed to an endless conveyor-belt which is driven by a steam-engine. Having reached the top, the block is seized by workers and dragged along a gallery giving access to the ice-storage rooms through doors spaced along its entire length. At each door, a workman tackles the ice-blocks with hooks and pushes them into the waiting repositories. When stacking the blocks, care must be taken to ensure that a space of 2 to 4 inches is left between them for ventilation and water drainage. When an ice-chamber has been filled, the ice is covered with a layer of loose hay and the door is closed until the ice is needed. It is then loaded into barges which convey it to New York where it is sold at about 1 cent per pound.

The ice collected by this method on the Hudson River and the Canadian lakes is shipped mainly to the Caribbean, South Africa, India and Australia. However, during the short, mild winters of recent years, the quantities of ice thus collected have been inadequate to cope with the vastly increasing demand—mainly because it takes too much time to cut, or rather tear, the ice into blocks.

This situation may be remedied by the steam-engine recently invented by Chauncy A. Sager, an American engineer. Moving along at a slow, steady pace, this engine saws the ice into blocks of suitable dimensions. The circular saw that cuts the ice lengthwise is fitted to a long lever at the end of the vehicle and connected to the driving-shaft by an endless belt. A pair of steel bars attached to the side of the engine carries a second circular saw, perpendicular to the first one. This saw is used for making cross-cuts. To enable the vehicle to move forward on the slippery ice, its wheels are fitted with sharp prongs.

By enlisting the aid of modern steam-power, the growing demand for ice can now be satisfied. When the icy winds sweep over the frozen plains, it is an odd spectacle to see people, working with frenzied haste and with the aid of steam-engines emitting large clouds of smoke, busily collecting the ice which, after many months have elapsed, will be sent everywhere to bring coolness and refreshment during the summer heat.

AN ARTIFICIAL SKATING-RINK

What is often denied to us by nature—a winter bringing ice for skating—will be provided by art, or rather science, for in the near future La Gran Plaza de Toros in the Rue Pergolèse in Paris will be transformed into an ice palace where a circular skating-rink more than 22,500 square feet in area will provide continuous facilities for ice-sport.

How is this artificial skating-rink with its diameter of nearly 200 feet installed and kept up despite the sometimes high temperatures which prevail outdoors? By evaporating liquid ammonia and using the cold thus generated for making ice. The area intended for the skating-rink is levelled and covered with a layer of concrete. On this foundation is laid, in numerous coils and loops, a network of fourteen iron pipes, each some 1,200 yards long, and having an inside diameter of $1\frac{3}{8}$ inches and a total length of 10 miles. Ammonia is piped through this system and the gas thus released is compressed and liquefied again by enormous pumps driven by three steam-engines delivering a total of 120 horsepower.

The main problem in keeping the rink in good repair lies in maintaining the tightness of the many joints which are unavoidable in a set of pipes of this length. And yet, this

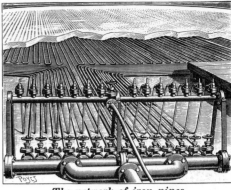

The network of iron pipes of the skating-rink [1890]

requirement is of primary importance. Should a leak occur in one of the pipes, the water surrounding it will come into contact with the ammonia gas which will immediately dissolve in the water. Due to the atmospheric pressure, the 3,500-foot-long coil in which the leak or burst has occurred will be entirely filled with water and thus be put out of action.

It is to be expected that such undesirable stoppages will occasionally interfere with the operation of the ice palace in the French capital.

Miller & McClellan's improved awning [1869]

MILKING MACHINE

An American, Edward M. Knoffin, has invented a milking-machine which, he claims, can milk a cow more rapidly, easily and hygienically than can be done by hand. Requiring no skill to manipulate, the device eliminates loss of milk and contamination, while it may also be employed when the cow is suffering from sore teats.

The apparatus consists of a glass globe A, sufficiently large to contain the maximum amount of milk yielded by one cow, a metal cover B, hermetically sealed by a pivoted bar and thumbscrew and equipped with spigots over which a number of rubber tubes have been fitted. One of the tubes is connected to an air-pump, the other terminates in metal tips that can be inserted into the cow's teats. A handle for carrying the glass globe is fitted to the cover, and may also be used for fastening the device to the cow's body by means of straps.

The cow-milker is extremely simple to operate. By turning on the suction-pump, the air in the globe is rarefied; the suction-valve having been closed, the metal tips are inserted in the animal's teats, the valves in the adapters between the tips and the tubes are opened, and the milk flows into the globe. A patent was issued for this useful invention on 3rd October 1876.

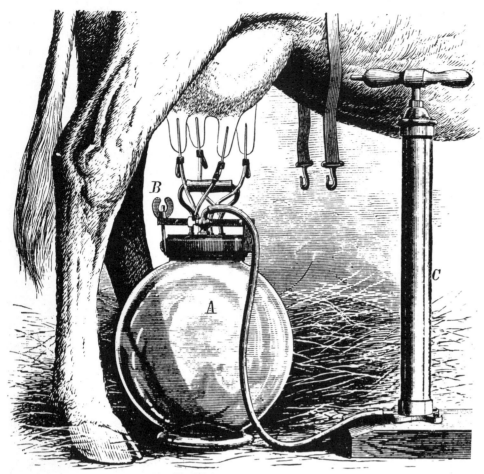

Knollin's cow milker [1877]

NEW DOMESTIC MOTOR

The inventor of the device which we present not only employs the hitherto wasted female power to oscillate a cradle, but at one and the same time to vibrate the dasher of a churn. By this means, it will be observed, the hands of the fair operator are left free for darning stockings, sewing, or other light work while the entire individual is completely utilized. Fathers of large families of girls, Mormons, and others blessed with a superabundance of the gentler sex, are thus afforded an effective method of diverting the latent feminine energy, usually manifested in the pursuit of novels, beaux, embroidery, opera-boxes, and bonnets, into channels of useful and profitable labour.

The apparatus consists of a lever A suspended from the ceiling or other suitable support from a swivelled hook and staple. In the extremities of the lever A are formed slots through which pass bolts and nuts which secure the adjustable arms B. To the eyes of the bolts are attached the end of two ropes, which pass round double guide pulleys fastened to the floor and then to two single pulleys, arranged one beneath the forward and the other beneath the rear part of the rocking-chair. The ends of the ropes are secured, as shewn, to the rungs of the latter.

Near the extremities of the arms B sliding weights are placed, by moving which the lever can be properly balanced. Just inside the weights is secured on one arm the dasher of the churn, and at the other a cord communicating with a cradle rocker. As the chair

New domestic motor [1873]

A steam-engine for domestic purposes [1884]

A STEAM-ENGINE FOR DOMESTIC PURPOSES

labour', and one in which she is not likely to be disturbed or encounter competition from the other sex.

A French mechanical engineer named Daussin has constructed a new type of steam-engine: to prepare it for operation, one need only place it upon an ordinary burning kitchen-stove. Thus the stove, which heats the working-man's kitchen and on which he cooks his humble fare, can also be used as a prime mover for machinery such as a sewing-machine.

This steam-engine consists of the following components: (1) an oscillating cylinder with a flywheel which also acts as the driving wheel; (2) a boiler with a capacity of only 2 pints, in which pipe coils form a heating surface of 365 square inches; (3) a hollow column integral with the boiler, acting as steam-dome and containing: (4) an automatic boiler-feeding device linked to the flywheel by a mechanical coupling, which ensures that an

is oscillated motion is communicated to the lever, and thence to both cradle and churn.

Necessarily this device may be put to a great variety of applications, and may supply motive power for washing-machines, wringers and other articles of household use, as well as for churns and cradles. At all events it opens a new field for 'woman's

Printing press driven by solar energy [1882]

Edison's phonomotor [1878]

adequate supply of fresh water, fed by gravity from a water-tank, is piped into the boiler under pressure by a system of valves operated by the head of steam generated by the machine.

Thanks to the fairly large heating surface and the small amount of water used, the boiler is capable of generating an adequate supply of steam in a matter of minutes. For example, if linked to a sewing-machine the engine can drive it at variable speeds which the housewife controls by means of a foot-pedal and a braking-disc connected to a helically wound spring and a counterweight.

PRINTING PRESS DRIVEN BY THE HEAT OF THE SUN'S RAYS

Recently, on 6th August 1882, Monsieur Abel Pifre, a French engineer, demonstrated the solar engine invented by him at a meeting of the *Union Française de la Jeunesse* held at the Jardin des Tuileries in Paris. It consists of a concave mirror $3\frac{1}{2}$ metres in diameter, in the focus of which there is placed a cylindrical steam-boiler equipped with a safety valve. The steam generated by the reflected sun-rays actuates a small vertical engine of $\frac{2}{5}$ horse power driving a Marioni type printing-press.

Although the sun lacked power and the sky was frequently overcast, the press operated continuously from 1.00 p.m. to 5.30 p.m. turning out an average of five hundred copies per hour of a journal which was especially made up for the occasion and appropriately

Combined stool and bustle [1887]

158

called *Soleil-Journal*. Previously Pifre had demonstrated that 50 litres of water could be brought to the boil in less than fifty minutes, after which the pressure of the steam increased 1 atmosphere every eight minutes. There is little doubt that such a solar engine will be a boon to the population of hot areas which so often suffer from a shortage of fuel.

EDISON'S PHONOMOTOR

No one will deny that the human voice is capable of producing an abundance of power, but this has never yet been harnessed to any practical use. In his experiments with the telephone and the phonograph, Mr Edison discovered that the sound of the human voice harboured a considerable amount of energy which could be converted into driving power and this led him to begin experimenting with the phonomotor which he himself has developed.

This apparatus has the same mouthpiece and diaphragm as the phonograph. The sound vibrations of the human voice cause the diaphragm to vibrate and these mechanical vibrations are then transmitted via a metal spring, one end of which is clamped in flexible rubber, to a wheel with a rough surface which shares a common shaft with a heavy flywheel. Speech sounds cause the flywheel to rotate, and if loud sounds are made into the mouthpiece over a period of time, considerable force is required to stop the flywheel as it spins round.

Mr Edison claims that a hole can be bored in a wooden wall by using his phonomotor, but we do not consider that such an application is of any importance since we are acquainted with voices which can perform this task without the intermediary of this complicated mechanism.

COMFORT AND FASHION HAPPILY UNITED

So long as it is the fashion for ladies to wear bustles of the pronounced amplitude now favoured by so many of the fair sex, we do not see why the fact may not be turned to advantage to build into the bustle a device calculated to make it convenient for the wearer frequently to rest from the fatigue of long standing and walking. We are curious to know what other ingenious devices the inventors hold in store for making further use of this space which has been neglected up to now.

HOW TO WEAR A LIFE-JACKET

It is imperative to show the passengers on board ship how a life-preserver should be

Delhommer's life preserver exhibitor [1880]

worn in times of emergency. The duty to demonstrate this adequately is often sadly neglected, a fact which may have fatal consequences should the ship come to grief. An American, Mr C. C. Delhommer, has now devised a water-tank made in the form of a human figure, and fitted with a life-preserver in the position in which it should be worn. As all passengers must of necessity make regular use of the water-tank to obtain drinking-water, they will repeatedly see how the life-jacket should be applied.

HAT-CONFORMATOR

Anyone who has ever tried to buy a new, stiff hat, be it of silk or felt, knows that even if the exact size has been established it is necessary to try on at least a dozen hats before one which fits pleasantly and comfortably is found. And even that is not always possible, so that one has no alternative but to buy a hat which pinches in some places, and augurs of future headaches.

To do away with such problems, a Frenchman, Allie of Paris, has now constructed a hat-conformator. Assembled from no fewer than 610 components, the apparatus is placed on the head like a hat and completely adapts itself to the shape of the skull. The astonished client hears a soft click over his head, and a moment later the hatter opens a cover in the device and takes out a diagram pricked on a piece of paper. This diagram and the apparatus can be used to give the hat the ideal shape to suit the customer.

The hat conformator [1879]

NAVIGATING IN FOGGY WEATHER

The serious collision between the steamers *Narragan* and *Stonington* due to fog has, once again, focused attention on the grave risks involved in navigating under foggy conditions. The fog-horn is the only means of indicating impending danger but it is difficult accurately to locate the direction from which the sound of the horn approaches us. Professor A. M. Mayer has now devised an apparatus to increase by artificial means the distance between our two auditory organs to enable us to find the direction of the sound with greater accuracy. The construction of the device will be apparent from the accompanying engraving.

How the hat conformator is used [1879]

Hamilton's improved gate [1884]

Forster's umbrella support [1888]

Professor Mayer's topophone [1880]

MOUSTACHE GUARD

In these days, when an unshaved lip is the rule, except among clergymen, it is hardly necessary to dwell upon the advantages or disadvantages of the moustache. Suffice it to say that, in our changeable climate, physicians are agreed that it conduces to health; and the inconvenience it offers, to the

Bartine's sunshade hat [1890]

Moustache guard [1872]

161

Each tone is a hole in the scroll [1879]

The nose has long been employed to support eyeglasses and spectacles. It is said it was once employed by a celebrated musician to execute a note inserted by an ingenious joker in a piece of music, the exigencies of which extended the hands to the ends of the keyboard, while the note in question required the manipulation of a key in the middle. Surely the nose, after having performed such a feat, must be equal to the keeping of one's moustache out of one's mush and milk, when provided with a proper instrument for the purpose.

Such an instrument is provided in Mr Randolph's invention. It is a curved plate of hard rubber, or other suitable material, adapted to the form of the upper lip, so that, being suspended in front thereof, the flange will take under the moustache, and hold it so as not to interfere with eating and drinking. Kissing, although not claimed in the patent, might perhaps also be rendered more easy and satisfactory by its use.

The plate has two curved prongs with rounded edges, so as not to injure the parts with which they come in contact, and adapted to enter the nostrils and suspend the plate from the thick part of the nasal septum, by grasping the latter, the prongs being inserted at the front of the septum, and pressed backward till they get a good hold. The moustache is thus held, as shown in the engraving, with the attendant advantages above set forth.

Gally's autophone or self-playing musical instrument [1879]

imbibation of the various fluids with which the human animal regales himself, has not been found sufficient to destroy the favour with which this popular hirsute appendage is regarded. In fact, it may be questioned whether it is not looked on with feelings of envy by certain strong-minded individuals of the sex whose faces Nature has denied the manly attribute of beard.

The man who has invented a means, whereby those 'bearded like the pard' may sip their wines, mixed drinks, and the milder beverages which 'cheer but not inebriate,' may justly be ranked in the long list of the eminent benefactors of mankind; and in virtue of his having conferred this inestimable boon upon mustachioed humanity, we therefore record, among the latest and brightest of these benefactors, the name of Eli J. F. Randolph of New York, who patented (20th February 1872) through the Scientific American Patent Agency, the device which is illustrated on p. 161.

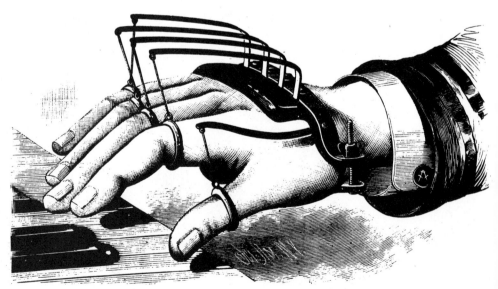

Atkins' finger-supporting device [1881]

FINGER-SUPPORTING DEVICE

Mr Benjamin Atkins of Cincinnati, Ohio, recently took out a patent which concerns an instrument designed to hold the fingers of a piano-player in a horizontal position as far as the second joint. The purpose of the new invention is to improve the player's touch and promote more rapid fingering.

A NEW VIOLIN

Professor Bruno Wollenhaupt of New York City has commissioned the manufacture of a new violin with twelve additional strings covering an octave in semitones. Fitted within the body of the instrument these will resonate in unison with the strings which are bowed in the usual way, thus greatly enriching the tone and volume of the instrument. With a slight movement of his chin, the player can operate and control a damping device which completely eliminates the resonance of the auxiliary strings. During a demonstration recently held in Berlin, Professor Joachim,

The Wollenhaupt violin [1895]

king of violinists, was lavish in his praise of the new instrument. Resonance strings may also be fitted to violas, violoncellos and double-basses.

THE 'CELLO PIANO

The most delightful musical instruments—those which make the greatest appeal to the soul by virtue of their similarity to the human voice—are without a doubt the violin, viola and the 'cello. They are more highly esteemed than the piano, for they allow the artist to make the notes himself in pitch and sound and volume. Unfortunately, stringed instruments are the most difficult of all to play and many pupils never succeed in mastering them since it is so difficult to estimate the proper pitch and correct it rapidly. The accuracy of the note is, of course, determined by the position of the fingers of the left hand which must coincide as accurately as possible with the desired mathematical division of the freely vibrating strings.

Reflecting on all this, a music teacher, Mr de Vlaminck by name, struck upon the idea of combining the melodiousness and sensitivity of tone of the string instruments with the tonal accuracy of such keyboard instruments as the piano and harmonium which is not dependent on the player.

After many experiments Mr de Vlaminck has finally met with success, and he has been able to take out a patent on a contrivance which is suitable for every stringed instrument in a quartet. In this new system the left hand of the performer is replaced by a mechanism which is brought into action by the keys of a piano keyboard.

Thus one plays as on a piano with the left hand, while simultaneously playing the violin, viola or 'cello with the right hand. The usual effects can be obtained with the bow, and the keyboard provides the accuracy of tone which has been permanently set for once and all. The keys are so perfectly linked to the hammers resting on the string that it is even possible to produce that slight tremolo we associate with playing 'with feeling.'

The inventor has constructed two types of instrument: the 'cello piano and the viola piano, both of which are shewn in the illustration. These are real musical instruments which are certain to enjoy widespread success. They have already been used for playing compositions by Beethoven and Haydn, and even to render the *Poet and Peasant* Overture.

The 'cello piano and the viola piano [1893]

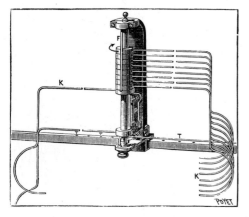

AUTOMATIC PAGE-TURNER

Augustin Lajarrige, a mechanic from Marseilles, has devised an apparatus that will turn sheets of music automatically, i.e. without any help from others or without the performer himself requiring to use his hands when the device is fitted to a piano, for example, the player need only move lever L with his knee from left to right, and the leaf will turn. As will readily be understood it is no easy matter to make an apparatus such as this and the construction is fairly complicated.

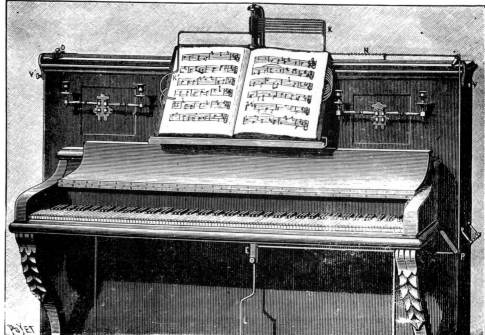

Automatic page-turner [1887]

TYPEWRITERS

Of all the noble arts, that of writing is surely the most frequently pursued. Any person who lays claim to a certain education nowadays must be more or less proficient in the calligraphic art. The fact that this proficiency is not very easily acquired, is proved by the six-year training course followed by each of us to develop a clear and steady hand in the pursuit of which many still fail to achieve satisfactory results.

For that reason, and to increase the speed of writing, a few inventors held the view that the old method of drawing letters by hand should be abandoned in favour of mechanical reproduction of already existing letters. The first useful writing-machine was constructed in 1865 by an American named Sholes. In 1877 a typewriter which gave excellent service was devised by an American engineer called Remington, but it was not until the end of the 1880s that the usefulness of the typewriter was generally recognised, and only then did people cease to regard it as an aid solely for the myopic and sufferers from writer's cramp. Now that it is no longer a rare occurrence for merchants or private persons to purchase a writing-machine to enable them to deal with their correspondence in a quicker and better way, we consider it expedient to give a survey of the most important of these machines.

Remington's typewriter of 1877 had a keyboard of forty-four keys, arranged in what was regarded by Remington as the most convenient manner. Consequently, the keys for the letters q, w, e, r, t, y, u, i, o, p are placed in one row. Only capital letters can be printed with it. An English lady who demonstrated this machine in Paris some time ago achieved a writing speed of more than ninety letters per minute, i.e. more than twice the speed attained in writing by hand. The relative expensiveness of Remington's and other similar constructions has induced Mr Herrington to create a device which, although

Sholes' typewriter [1872]

The 'Remington' typewriter [1891]

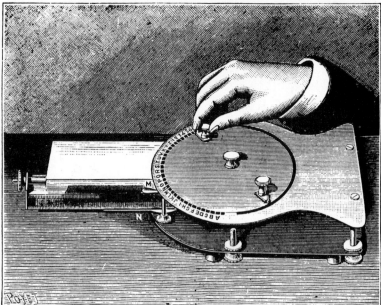

Eggis' 'Vélographe' [1891]

less perfect, is likely to give satisfactory results once the user has familiarised himself with its operation.

Its main component is a wheel which can be held in the hand and rotated about an axis. The twenty-six letters of the alphabet, the numbers and punctuation signs have been carved round its circumference. When the wheel is rotated, these signs come in contact with an ink-cylinder above them which provides them with printing-ink. The wheel can easily be moved up and down, and forward and backward. In order to print what is to be written, the wheel is rotated until the desired letter is above the paper and then pressed down through a slot in the guiding-rod so that the letter touches the paper. After printing, the wheel is raised by the action of a spring. To print the following letter, the wheel can then be moved one tooth to the right along the guiding-rod under the writing-frame. A button is pressed with the left hand to enable it to do this—see the figure. It is also shewn how a frame holds

the paper to be written upon. To write on the next line, the paper is moved the appropriate length upwards. It must be regarded as a drawback that this is not done mechanically. But it should not be forgotten that the whole machine costs no more than some twenty shillings.

Another very simple writing-machine is the Vélographe (or rapid writer) of Mr M. Eggis. The name is somewhat pretentious for this typewriter cannot be worked at a very high speed. The characters have been placed on the lower side of a revolving disc: the capitals and figures on one half, the lower-case letters and punctuation marks on the other. By moving one of the two buttons on the disc until the indicator is opposite the desired letter on the edge of the circle and then depressing the button, the letter, figure or stop is printed on the paper wrapped round a cylinder below the disc. Since the disc carrying the type is easily interchangeable, various sorts of characters can be printed with the same instrument. With this type-

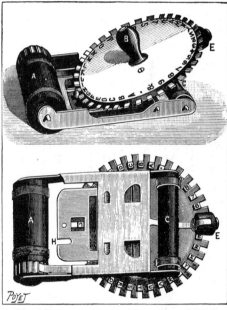

The Miniature Pocket Typewriter [1891]

writer by Eggis, one can do all that is desired, provided one has the time and patience for it.

The Miniature Pocket Type Writer is a typing-machine of pocket-size and, as such, probably the smallest in the world. The rubber letters have been placed on the underside of a metal disc having a handle in the centre. If the machine is to be used, it is placed on the paper at the place to be typed upon and held fast by the left hand, supported by cylinder A. Lever B is rotated until the type to be printed is brought above a certain aperture. Now, by gently depressing lever B a pin is inserted, causing the letter-type— inked in the meantime by cylinder E—to touch the paper underneath and leave its impression on it. As regards speed, this simple device will undoubtedly be no match for the larger machines but it should be realised that it costs about fifty times less than these.

In the Columbia typewriter, there is a

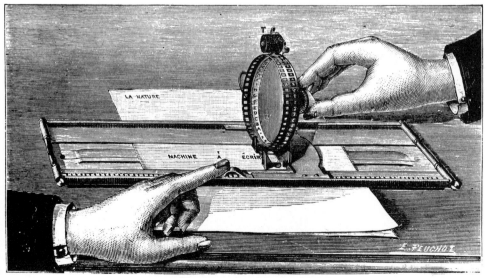

Herrington's typewriter [1891]

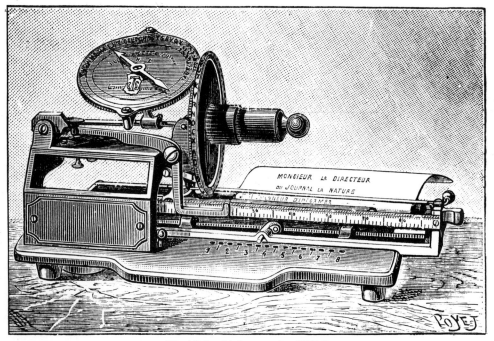

The 'Columbia' typewriter [1891]

every civilised family will have, along with the sewing-machine, its own typewriter. If the next generation familiarise themselves with the machine in childhood, they may at a later age achieve a typing-speed which is inconceivable now. But who knows, perhaps at that time, the ideal typewriter will have been invented which, having ears like a phonograph, will accurately reproduce in typescript the words spoken into it.

A SHORTHAND TYPEWRITER

This is a typewriter built especially for rapid work; simple enough to be very strong and small enough to be light, portable and noiseless. In fact, it is not much larger or heavier than a pair of opera-glasses. Speed is gained by arranging the keys and type so that every letter on the keyboard can be printed at one time without shifting the hands, all the most frequently used letters being duplicated. Thus in writing the word 'start,' the 'sta' would be struck with the left hand and the 'rt' with the right hand simultaneously, the entire word being printed at one stroke, after which the machine automatically draws the paper forward and is ready for the next word to be printed, so that it requires no more strokes of this kind to print a whole sentence on the Anderson Shorthand Typewriter than it would merely to strike the space key for making spaces between the same words on an ordinary typewriter.

This arrangement of the keyboard restricts the number of keys and necessitates the omission of the less frequently used letters of the alphabet. These omitted letters are represented by combinations of those the machine prints, and as soon as the list or code of cipher letters is memorised, the learner has a complete alphabet at his service and can begin practising for speed. Six weeks' practice will, it is said, give a speed of about 100 words per minute. No knowledge of stenography is required; there is nothing to learn except the list of cipher letters.

type-disc which can be moved round its axis in a vertical plane by means of a knob placed in the centre. In this machine, a gear-train links the type-disc with an indicator-arrow that moves across a slanted dial bearing all the letters and punctuation marks. To print the desired letter, all one has to do is to depress the knob of the type-disc which is possible beause the entire mechanism can rotate round a spindle. After this, the paper which is sandwiched between two platens or cylinders is moved along a distance equal to the width of one letter. A speciality of this Columbia machine is that for wider letters such as *m* and *w*, this interval is larger than for the *e* and *a*, while it is smallest for letters such as *l* and *i*. This makes the typescript of this instrument look more like letterpress.

One of the most ingenious writing-machines is the Grandall, in which the eighty-four different characters have been embossed in six rows on a metal cylinder or 'type sleeve' which, when a specific key is depressed, moves up or down, or turns round as may be necessary to ensure that the corresponding type is printed. By releasing the key, the type-cylinder is returned to its rest position by a helical spring.

One of the finest typing-machines available is undoubtedly the Hammond, the types of which are placed on two segments of a cylindrical surface; its carriage is a marvel of ingenuity. At the end of the line, the platen is automatically moved up one line-space; this line-space is adjustable.

In view of the high degree of perfection now attained by the best typewriters, and the fact that efforts will still be made to achieve further improvement, it may safely be assumed that in the not-too-distant future

AN ELECTRIC STENOGRAPHIC PEN

Mr Augustus S. Cooper of Santa Barbara, California, has invented a stenographic pen which writes by means of electric currents brought into operation by the left hand and the feet. One pole of the battery is connected to a metal plate on which is placed a specially prepared sheet of paper. The other pole is connected by ten conducting wires to ten metal pins at the end of the pen. In nine of these pins there is a push-button switch,

The 'Grandall' typewriter [1891]

166

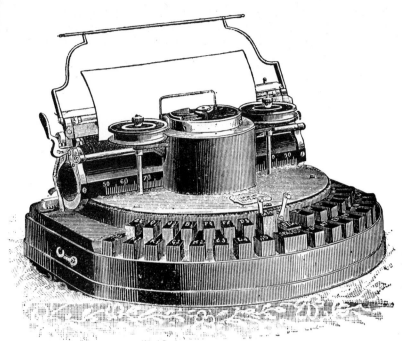

The 'Hammond' typewriter [1891]

The electric stenographic pen [1890]

comparable to a signalling key, with which the circuit can be opened or closed. Five of these are operated by the five fingers of the left hand, one by the palm of this hand, one by a sideways movement of the fingers, one by the left foot and one by the right foot. If one of the metal pins connected to the battery comes in contact with the specially prepared paper it becomes discoloured and a dot appears on it. Anyone experienced in operating the push-buttons moves the pen across the paper as he does so, and thus obtains a pattern of various combinations of lines from which the stenographed text can be read back.

A shorthand typewriter [1893]

MACHINE FOR WASHING DISHES

One of the largest restaurants in Paris has begun using an automatic dishwashing-machine invented by Monsieur Eugène Daguin. It consists of a circular tank divided into two compartments, one for hot and one for cold running water. The dishes to be washed are held in eight artificial hands revolving round a central shaft. After passing through the hot bath with an undulating movement to wash off the grease, the dishes are vigorously brushed by two rotating brushes and dipped into a cold bath where they undergo the same undulating movement. Finally, the plates and dishes are taken out and placed in a draining- and drying-rack. The use of this machine constitutes no danger whatsoever either to man or dish!

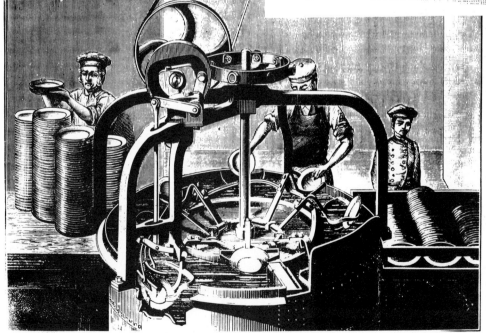

Machine for washing dishes [1885]

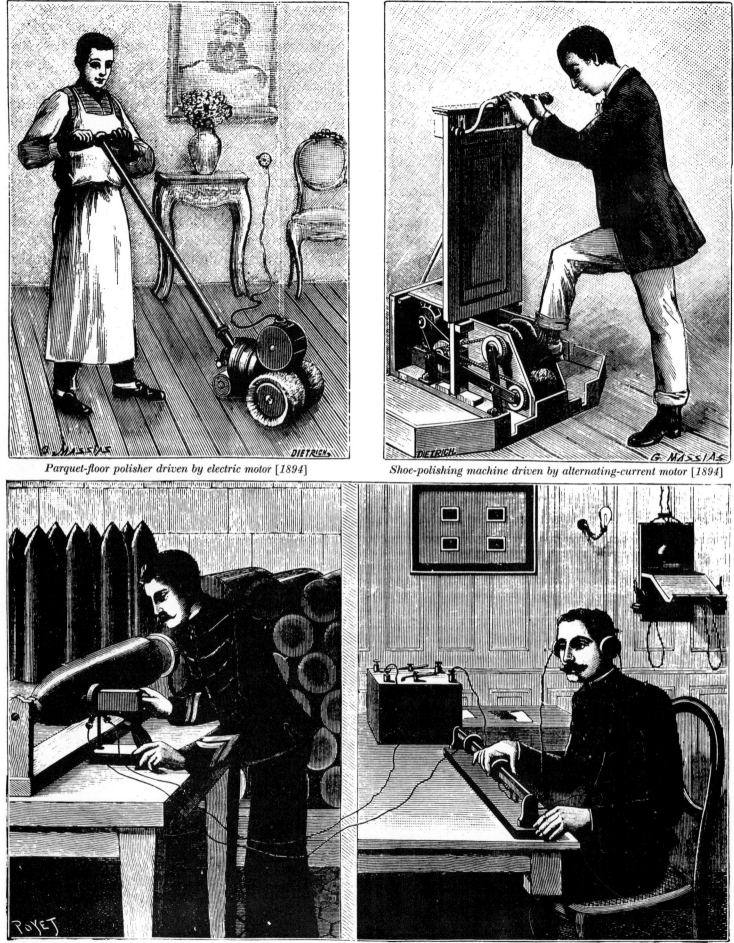

Parquet-floor polisher driven by electric motor [1894]

Shoe-polishing machine driven by alternating-current motor [1894]

The schizeophone and how it is used; left, the testing-room; right, the recording room [1895]

THE SCHIZEOPHONE

In 1890, the French captain Louis de Place invented a device to detect internal cracks or holes in metals. He called the apparatus 'schizeophone,' a contraction of two Greek words meaning 'crack' or 'split' and 'sound.' The device consists of a small hammer, a microphone, and a sound-meter. The hammer is moved to and fro by means of a simple mechanism such as a set of gears, or by hand, at a speed not exceeding three strokes per second in order to make the sound clearly audible. The hammer passes through a microphone composed of carbon prisms placed in a square or triangle on square blocks of carbon. The sound-meter connected to these elements is a calibrated wooden pole round which two wire coils have been placed, one fixed and the other sliding. The movable coil is connected to two telephones secured to the observer's head by a ribbon.

The operating principle of the apparatus is extremely simple. In one room, the metal being tested is struck with the hammer; in the other, the observer will hear, in his ear-telephones, a sound generated by the impulses induced by the fixed coil in the circuit of the movable coil connected to the telephones. The volume of the sound is dependent on the distance between the two coils. The observer now adjusts the distance so that a scarcely audible sound is heard in the telephones. Any increase in the sound volume will then be easily perceived.

As soon as the hammer hits a spot where a crack or cavity is present in the metal, this flaw will act as a sounding-board and reinforce the sound, thus permitting the observer with the telephone to identify the flaw whose

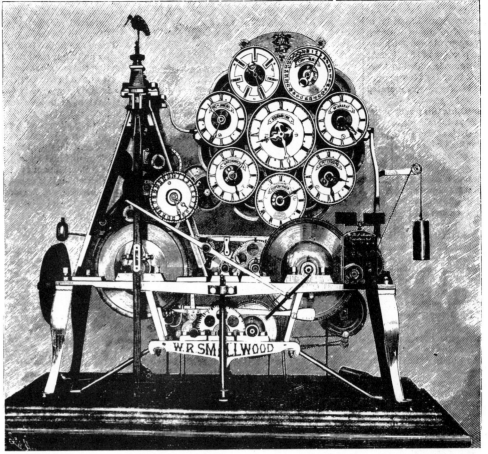

Smallwood's eight day sixteen-dial clock [1892]

presence he can then indicate by pressing a bell-button.

Since the frequent occurrence of fractures in railway lines is attributed to internal defects in the rails, the French railways are now having the materials for the iron track examined in their depots. The illustration shows the testing of artillery-shells. Cracks in the metal jacket of such shells are likely to be a source of serious hazard.

Delostal's electric match [1893]

AN ELECTRIC MATCH

This apparatus, invented by the Frenchman Monsieur Delostal, consists of a bell with an opening in the top into which a stick A is inserted. This stick functions as a match and at the bottom it terminates in a hollow perforated space filled with cotton. This comes out at the bottom of the apparatus into a chamber E which is filled with pieces of felt and a highly inflammable liquid—a mixture of alcohol and ether. The device is connected to a galvanic battery by means of two wires.

When the stick is pulled out, an electric spark ignites the cotton soaked in the inflammable fuel. In the space B it passes between two springs which close the chamber when the stick is drawn upwards. C is a funnel which opens at two places to permit the stick to pass through. A small watch-spring D on an insulated copper column, which receives current through wire F, produces an electric spark when the contact with the bottom of the stick is broken. The second wire is conductively connected to the earth of the device. When it is empty, the chamber can be refilled with liquid through an opening in the base.

New fire escape [1882]

Morell's floodway for warehouses [1875]

A STREAM-SPREADING WATER-NOZZLE

By shaping the nozzle of a fire-brigade hose in a special way and fitting it with radially mounted, adjustable levers, Mr Charles Oyston of Little Falls, N. Y., has succeeded in obtaining a series of tapered, diverging water-jets.

This results in a mist of fine spray which is said by the inventor to be more efficient in extinguishing fires than a solid column of water. By turning an adjustment ring on the nozzle, the fireman can produce the diverging jets of water shewn in our engraving.

LESCALE'S AUTOMATIC FIRE-ESCAPE

The modern multi-storey buildings with their frequently narrow passageways and exits should be equipped with an appropriate

A stream-spreading water-nozzle [1865]

Lescale's automatic fire escape [1878]

means of rapid escape for the occupants in case of fire. Mr John M. Lescale has now devised a life-saving appliance on wheels which may be kept in the house as a piece of furniture, to be moved to a balcony or window in the event of fire breaking out. After folding out the brackets of the apparatus, the person wishing to escape takes his seat in a ring formed by a length of cable with a ring and a hook, and attached to a cable which is wound on a reel. While descending, the user grips the rope connected to the braking

Pauly's fire apparatus [1893]

Electric fire-extinguisher [1898]

device with one hand. Only when this rope is pulled can the cable uncoil from the reel to permit descent. When the descent is completed another person may leave the building by using the second cable during which operation the first cable is automatically pulled back up. In this way, a large number of people can leave the burning building.

ELECTRIC FIRE-EXTINGUISHER

The provision of an electricity supply in a dwelling or office-block is useful not only for the purpose of lighting and for providing motive power to operate small household appliances, but may also serve as an effective expedient for extinguishing fires. The firm of Merryweather & Sons, of Greenwich, near London, has designed a mobile fire-extinguishing apparatus capable of holding more than 20 gallons of water. A pump driven by an electric motor connected to the electricity mains produces a powerful jet of water which will put out any fire breaking out in the home quickly and efficiently.

AN AUTOMATIC SCULPTURING MACHINE

Monsieur Delin, a Paris manufacturer of religious statues, has designed a sculpturing machine in which skilful use is made of electric motors. It consists of two sturdy columns united by an upper member and a base plate. In the centre, there is a vertical spindle provided with a carriage capable of moving up and down throughout the length of the spindle, actuated by an electric motor placed at the upper part. The carriage is equipped with two supports extending to the left and right in front of the statues shewn. They are provided with slides in which are placed the apparatus that serve for the work, *viz.* to the right, the pantograph operated by the workman in front of the model, and to the left the sculpturing machine. The two apparatus are capable of moving around the central spindle, and every motion at the ex-

tremity of the one is reproduced at the extremity of the other, as in every pantograph. The workman holds a wooden rod which follows the contours of the model, and the same motion is accordingly transmitted to the work to be shaped from the solid. This is effected by an electric drill placed at the extremity of the arm to the left. It sets in motion an auger bit that revolves with great velocity. When the machine is in operation, it suffices for the workman to bring the wooden rod near the model, as shewn in our figure when the auger bit approaches the piece of wood and cuts out a portion in such a way as to reproduce the model. The workman can likewise cause the carriage to rise or descend in order to effect the same work throughout the length of the statue.

———

COMBINED HORSE-POWER AND STABLE FLOOR

The engraving depicts a mechanism which enables a horse to clean its own stall, cut up its fodder finely, press out sugar-cane, drive a flour-mill, a maize-huller or a bellows, or can even pump water which may be used to irrigate the soil, for drinking or extinguishing fires. This mechanism is always ready to go into operation and can even be used with a cow or a bull. The horse can be allowed to run for hours without ever having to leave its stall and can be kept in good condition by this training. The horse power can, of course, be used for a number of other purposes such as the generation of electricity for lighting. A is an endless floor, B and C are the shafts whose rotation is transmitted to a central shaft, the drive belt can be tightened with D which is equipped with a wheel and a screw, and E is a counterweight which presses a brush against the endless floor and thus keeps it in a constant state of cleanliness.   [1880]

Automatic sculpturing machine [1894]

AUTOMATIC SCENT-DISPENSERS

For some time now the doors of shops, theatres, concert halls, etc. in Paris and other large cities have been equipped with metal bottles of various colours, carrying enamelled signs inviting passers-by and customers to spray themselves with some fine scent for ten centimes or some other coin. The mode of operation is quite interesting. The coin drops from slot A through a channel B on to platform C. If button D is pressed, the coin moves against cylinder F and pushes it into the hollow cylinder K where it serves as a plunger, causing a displacement of the channel in the vicinity of F. This slight movement causes a small amount of air to escape through the space cleared above the channel and to enter tube T which communicates with the open air. Simultaneously, a few drops of scent flow from the receptacle L through the cleared channel into tube H and from there into the open. When button D is released the cogs P which are placed in such a manner that they have to travel a certain distance before they engage the upstanding rims of cylinder F, release the coin and allow it to drop to the bottom of the metal bottle. The quantity of scent dispensed is regulated by the position of screw V.

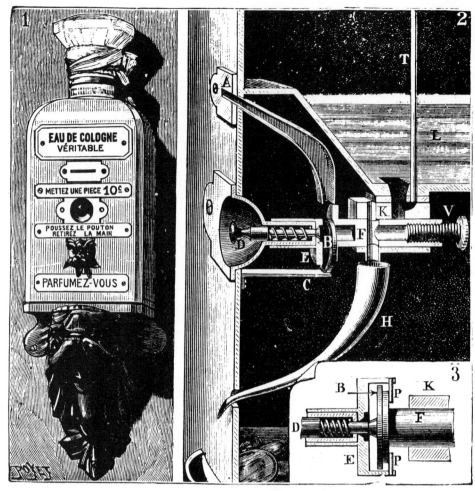

Automatic scent-dispenser [1895]

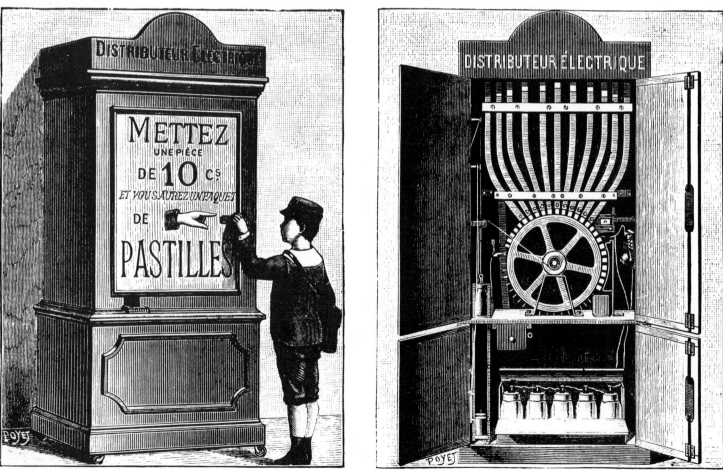

Chocolate-dispensing machine. Interior of the electric automaton [1887]

AUTOMATONS

At the present time, we are being inundated with automatons. If this continues, the time will come when all the arts and crafts will be performed by machines which, at the cost of a coin, be it large or small, will be at everybody's service. At this point, human hands will be necessary only for producing the automatons, unless of course automatons are invented which are capable of performing this work too. One cannot enter a public place nowadays without seeing a weighing-machine, a chocolate-machine and frequently also a penny-in-the-slot-machine rendering some popular waltz at the cost of a copper or two.

At the World Exhibition now being held in Amsterdam, these ingenious machines are present in a wide variety of shapes and forms, each more ingenious than the next. There are automatons for beverages dispensing refreshments for a few coins, hens laying tin eggs filled with sweets and, as befits a chicken, cackling loudly as the eggs emerge to announce that they are on the way. There

are dainty automatic misses who offer you a choice of chocolates on a tray after you have sacrificed a penny. But the chief object of the visitors' admiration is the slot-machine that produces within the brief space of three

Weighing-machine with automatic indicator [1886]

minutes a tolerable likeness of whoever cares to pay a shilling for his portrait. This is such an amusing and surprising novelty that few visitors will be able to resist the temptation of sitting down in front of this machine in order to take home their merriest exhibition face, neatly packed in a case. For one need do no more than sit down upon the seat and put two sixpenny pieces in the slot. Rumbling sounds are then heard in the depths of the machine, a pointer is observed to move a dial, and when it has made one complete revolution the handsomely framed picture falls into your lap. After nightfall, magnesium flashlight is used.

Ferrer's photographic automaton [1895]

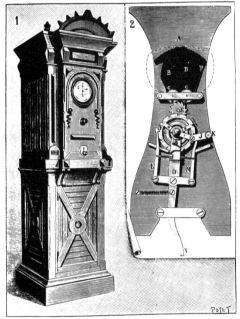
Interior of Ferrer's automaton [1895]

Automatic hot-water dispenser in Paris [1893]

175

Edison and his phonograph [1878]

EDISON'S TALKING MACHINE

Ten years ago, in December 1877, a young man entered the editorial office of the *Scientific American* and showed the people present there a simple machine of his own design which he had completed only the day before. Having given a brief explanation of his invention, the visitor turned a crank and, to the amazement of those present, the apparatus said: 'Good morning. How do you do? How do you like the phonograph?'

Edison's phonograph [1878]

And so the instrument spoke for itself, introducing itself as the 'phonograph' or talking-machine.

The young man was Thomas Alva Edison, already well known for various successful inventions. The rumours about this wonderful machine also reached Europe. There, however, no one believed in the possibility of an apparatus capable of recording the spoken word and reproducing it again; 'the usual American humbug' was the consensus of opinion. In the meantime, Edison had adapted his machine to the requirements of everyday use, and sent his representative,

Registration of a cornet solo by the phonograph [1889]

Puskas, with an improved model to Paris where it was presented to the members of the French Académie des Sciences on 11th March 1877. The learned society spent an unusually enjoyable evening with it.

The members had turned up in large numbers, and every seat in the public gallery was taken. The meeting having been opened, Mr Puskas placed himself directly in front of the apparatus and said in a loud voice and with a strong English accent: 'Le phonographe présente ses compliments à l'Académie des Sciences.' A few moments later, amid a profound silence, the audience heard the apparatus repeat the same words with all the inflections of the speaker's voice, including the English accent. The success was such that the distinguished company burst into spontaneous applause, and requested him to repeat the experiment. Thereupon, the American spoke a new sentence into the apparatus. This time it contained both question and answer, and the machine repeated in a humorous way: 'Môssieu phonographe, parlez-vô français?' 'Oui Môssieu.' These words were clearly understood by everyone, but were spoken at a lower pitch than that of the original voice which completely baffled the audience. They were, it seemed, convinced that some mystery was afoot. And yet the lowness of the pitch was most likely due alone to the crank's being turned at insufficient speed. Now, the President of the Académie was requested to speak into the machine. In a loud clear voice, he said: 'L'Académie remercie Monsieur Edison de son intéressante communication.'

Moreover, in the proceedings of this notable meeting it is recorded that from then on, there was complete uproar so that it was only with the utmost effort that the Chairman succeeded in restoring normal order.

Listening to the phonograph at the Paris exhibition [*1889*]

This presentation in the French Académie contributed greatly to the rapid spread of the fame of the new machine, causing many people to put it to the test themselves or at least to attend the demonstrations which were given. At each exhibition the apparatus was loudly acclaimed by large audiences.

A MACHINE OF MANY USES

'What do you think the phonograph can be used for?' a visitor once asked Edison. 'For a multitude of purposes,' Edison replied.

'1. It will replace the stenographers. When someone has many letters to write, he will dictate them to the phonograph, and thereupon dispatch the tinfoil with the recording to his correspondents who have but to attach it to their own phonograph to hear whatever they may have to answer or act upon. Such letters, addressed to persons who do not own a phonograph, will be copied by the employee of a phonographic agency.

'2. An accomplished reader will speak one of Dickens's short stories into the phonograph. The entire story can be recorded on a fairly small tinfoil. By a simple operation, this foil may be multiplied millions of times. The family are sitting at the tea-table. They hear the short story, read to perfection by the skilled elocutionist already referred to.

'3. The phonograph will sing with the voices of a Patti and a Kellogg; at home, grand opera may be enjoyed every night at the cost of a few pence.

'4. The phonograph may also be used as a composer of music. Whenever a beloved aria has been sung, it may be stored by the machine and reproduced on countless occasions. By playing it back in reverse, an entirely new aria can be produced.

'5. It can read to the blind, or to a block-head who never learned to read.

'6. The phonograph can teach languages. It can teach children to pronounce the individual syllables. Had Stanley had a phonograph at his disposal, he might have collected and stored for the world's scholars all the dialects of Central Africa.

'7. It can be used to make toys talk. Dolls will be given all kinds of voices. Clocks will tell the right time.

'8. It will be used by actors to help them memorise the correct enunciation and delivery of their lines.

'9. The phonograph will reproduce the voices of absent relatives or those long since dead as well as of our loved ones far away. It will record the voices of children and the last words of the dying.

'10. It will become an addition to the telephone and store important conversations.'

In the ten years of its existence, the phonograph failed to fulfil any of these high expectations. To his bitter disappointment, Edison's best-loved invention was not put to any practical use. The phonograph was not to get beyond the scientist's laboratory to which it was relegated as a curiosity, and it is almost impossible to believe that in his time, Edison harboured such excessive expectations of his talking-machine.

Ten years elapsed in which no improvement was made on the phonograph. Edison himself, it was said, had no time to perfect his invention; he was preoccupied with other problems concerning electric lighting and heating, with his work on the telegraph and telephone, etc., which were to bring substantial fortunes to so many. However, a man like Edison could not be expected entirely to abandon an idea once conceived; nor would he confess his inability to accomplish what he believed to have achieved—or nearly achieved—already. It is therefore hardly surprising that Edison should once again take up his favourite machine in an attempt to perfect it.

As an unusual New Year's gift to its readers, the *Scientific American* in its issue of 31st December 1887 gave an account of the new Edison phonograph showing, among other things, the following improvements. The acoustic vibrations which are

Registration of a grand piano by the phonograph [1889]

Performance of the phonograph in the Trocadéro in Paris [1897]

French phonograph [1897]

Details of the needle and the cylinder [1897]

transformed into up-and-down mechanical vibrations by a diaphragm of gold-beater's parchment and a stylus attached to it, are no longer impressed upon a tinfoil wrapped round a cylinder, but on a hollow tube of hardened wax. These impressions may be removed to permit the wax surface to receive new sound impressions. A second alteration affects the cylinder which no longer moves along while rotating; it simply revolves round its axis, and now it is the mouthpiece with diaphragm and stylus which travels along the surface of the wax cylinder. The machine is no longer worked by hand but driven by an electric motor.

A staff reporter of the *Scientific American* has been present at a few experiments conducted with the new phonograph in Edison's laboratory. An article was read from a newspaper during the reporter's absence. Upon his return to the laboratory, he heard the article reproduced so clearly that he understood every single word although the names in it were unknown to him. Another proof of the degree of perfection achieved by the machine was the true rendering of whistles and whispers; every inflection of the human voice was faithfully reproduced.

The same reporter describes a few uses to which, according to Edison, the phonograph may be put. At legal inquiries, it could serve as an incorruptible witness; it would have only a single story to tell and cross-examination could not upset it. A lawyer or witness might talk as rapidly as he wished; each word would be indelibly recorded on the wax cylinder. Civil and military orders could be given by means of a phonograph cylinder. A typesetter could obtain his copy directly from the phonograph.

A book such as Dickens's *Nicholas Nickleby* could be recorded in its entirety on four 8-inch wax rolls 4 inches in diameter so that a book of this nature could be read to a large number of persons. The cylinders are extremly light and can be dispatched by post in boxes specially made for the purpose.

The phonograph will be put to interesting and popular use in making songs sung by famous singers more widely known, reproducing sermons and speeches, spreading the words of famous men and women, familiarising people with great music and playing back the calls of animals, etc. so that the owner of a phonograph may enjoy all these pleasures in his own home.

The necessary steps have been taken to produce the new phonograph on a large scale and, according to the American reporter, these machines are likely to become as widespread and indispensable as the sewing-machine or typewriter.

AN APPARATUS FOR INSCRIBING SPEECH SOUNDS

Monsieur Rousselot of Paris has constructed an apparatus in order to investigate the differences which exist between the dialects spoken in certain parts of France and the normal pronunciation of the language. For this purpose he employs five recording devices, each connected by gutta-percha tubes to a needle which inscribes five adjacent wavy lines on the blackened paper of a rotating drum, while a tuning-fork, which is kept continuously vibrating by an electric current, draws a sixth wavy line. The speed of rotation can thus be controlled in the most accurate possible way.

The first recording drum, $\frac{1}{2}$ inch in diameter, was placed with the diaphragm against one of the side walls of the thyroid cartilage (Adam's apple) and it registered the vibrations of the larynx while the subject was speaking. In order to record the movements of the tongue during speech, a second drum was placed with the diaphragm behind the subject's chin and secured by a band passing over his head. In this case, the movements of the needle mainly indicated the movements of the sub-lingual muscle. There was a separate recording device for each of the two lips. Its membrane was depressed whenever a highly sensitive lever was displaced by the movement of the lips.

Acting on a suggestion by Dr Rosapelli, in order to investigate the part played by the nose in speech, tiny gutta-percha globules were gently pushed into the nostrils and these transmitted to the needle on the recording drum with which they were connected the vibrations generated in the air of the nasal cavity when nasal sounds were made.

However ingenious in conception the apparatus used by Rousselot may be, it is nevertheless doubtful whether the results obtained with it will satisfy the expectations of its inventors. Speech is such a complicated

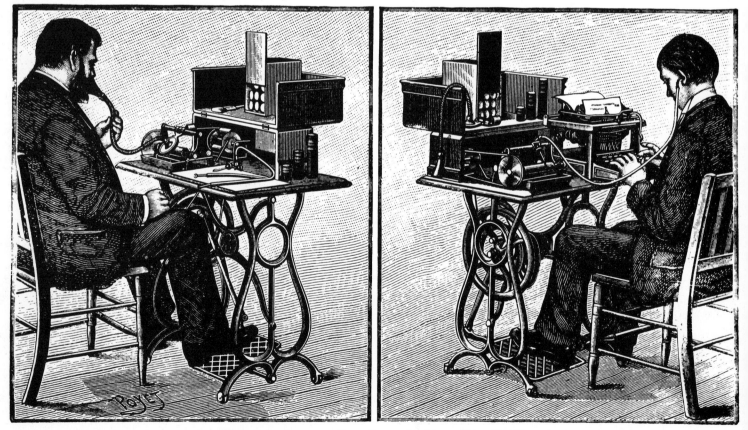

The graphophone recording—the graphophone talking [1889]

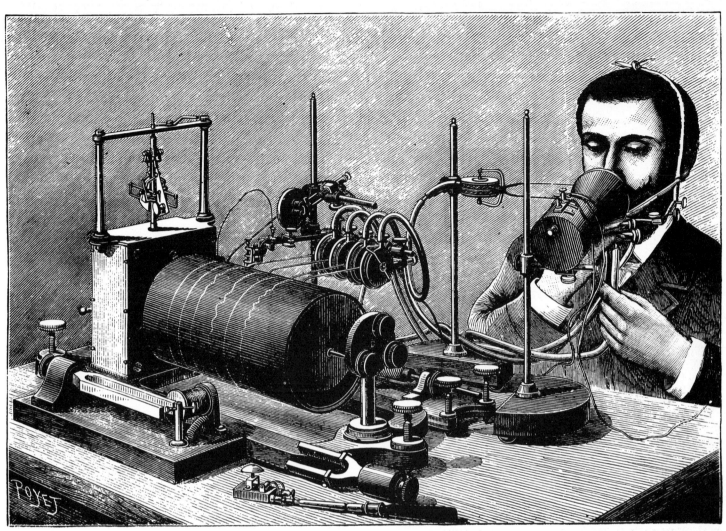

Rousselot's apparatus for inscribing speech sounds [1895]

process, influenced by so many other factors than those studied by Rousselot, that it is questionable whether a new Champollion has already risen up among us who is capable of deciphering the hieroglyphs inscribed upon the rotating cylinder.

———

TALKING WATCHES

In the first reports on the phonograph, invented by Edison in 1877, it was remarked that it would now be possible to produce timepieces capable of calling out the hours instead of indicating them by chimes. Instead of giving twelve successive peals or even saying 'cuckoo' twelve times in succession, the clock would call 'twelve o'clock,' 'quarter past twelve' and 'half past twelve,' etc. at the appropriate moments of the day.

Mr Sivan, a watchmaker of Geneva, appears to have succeeded in giving this temporary power of speech to an ordinary

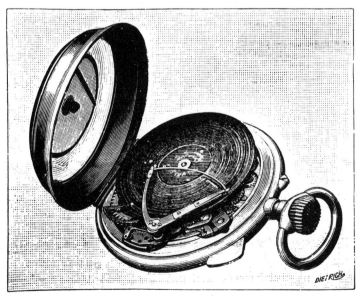

Sivan's talking watch, the mechanism and the phonograph disc [1895]

181

The manufactory of phonographic dolls [1894]

182

pocket-watch. It contains a phonograph disc made of vulcanised rubber having forty-eight grooves which correspond to the twelve hours and thirty-six quarters. If the button—which is similar to that of a repeating-watch—is pressed, the rubber disc starts rotating and a stylus, following the mounds and dales of the grooves, starts vibrating. It then transfers these vibrations to a membrane which converts them into sounds: 'twelve o'clock,' 'quarter past twelve,' and 'half past twelve' and so on, says the watch, reproducing a human voice.

A device such as this can be fitted to any clock. Indeed, Sivan has already manufactured alarm-clocks containing a talking disc which, at a specific time, calls out: 'Wake up!,' 'Get up!,' or, 'It is now time to get up!'

PHONOGRAPHIC DOLLS

The wizard of Orange City, New Jersey, does not forget the little ones although he is so greatly occupied with matters which concern their elders. For a long time now reports of importance to the children have been circulating to the effect that Thomas A. Edison has invented phonographic dolls.

'Just imagine!'—parents would say to their little daughters listening in wide-eyed wonder—'Just imagine! Dolls which can truly speak and do not merely gurgle and squeak, but say 'Mamma' and 'Papa' in a real human's voice, and tell fairy-tales and sing songs like real live boys and girls.'

And then the childish eyes would glisten and plump little hands would be clapped in glee. 'But how can that be?' and 'How is that done?' they would ask, for a child must always know the ins and outs of everything.

The answer is given in the well-known periodical, the *Scientific American*. The doll on the left looks such a clever creature that one might almost immediately think it was more than just an ordinary doll. And that is indeed the case. You have but to take off its clothes and open up the back and the speaking mechanism, the phonograph, will be revealed. This is shewn separately in the enlargement, bottom right. The largest cylinder is coated with wax and a sharp needle fitted to the membrane, or diaphragm, of the funnel comes in contact with it from above. If one speaks into the funnel the diaphragm begins to vibrate; the needle vibrates along with it and cuts the phonographic patterns in the wax in a spiral since the wax cylinder is driven forward, rotating in the direction of its axis.

Once the words spoken into the funnel have been recorded, the needle is raised slightly (so that it does not scratch the wax) and the cylinder is turned back to its original position by means of a key. Now the doll can begin to speak. A coiled spring presses the cylinder back and the needle, which has been lowered again, follows the line of the groove which has already been made in the wax.

Edison's phonographic doll—the built-in phonograph [1894]

This causes the needle to vibrate and it, in turn, sets the diaphragm of the funnel vibrating so that it reproduces the selfsame sounds which it had at first picked up. A small endless belt causes a flywheel lower down to rotate, thus ensuring a constant movement.

A part of Edison's factory is illustrated at the bottom left. Edison has no less than 500 people employed in manufacturing phonographs and half of them work in the doll department. Walking through the factory, one is filled with admiration for the order which prevails everywhere. Everything is done in the American way and the principle of the division of labour is most extensively applied. The tools and machines used here are of the very finest. There are standard measures for all the parts used in making the phonographs and all the components are carefully compared with these before they are given to the workers whose task it is to assemble the complete machine. This ensures that every part always fits properly in its place. About 500 talking dolls ready to play can be supplied every day. In the centre picture a female employee can be seen speaking the words on to the wax cylinders one by one.

The second set of illustrations shows a phonographic doll of French manufacture. This is fitted with a system of cog-wheels which is set in motion by pulling the knob A. This doll speaks French saying, for example: 'Je suis bien contente, maman m'a promis d'aller au théâtre, je vais entendre chanter, tra la, la, la, la,' and then it begins to sing a pretty little song or even to laugh and ends by saying: 'Merci, ma petite maman.' The wax cylinder can be taken out of the doll and be replaced by another one with a different recording; these recordings are made by young girls with clear, childish voices.

A PICTURE-BOOK WITH ANIMAL NOISES

This year many French children will receive a pleasant surprise in the form of a thick book full of splendid colour plates of animals. There are nine strings in the book. If they are pulled in turn the sounds corresponding to the pictures are heard one after the other: the crowing of a cockerel, the braying of a donkey, the baa-ing of a lamb, the bleating of a sheep, the twittering of young birds in their nest and the characteristic sound of the cuckoo, the billy-goat and some birds. These sounds are produced by tiny instruments concealed in the binding and consisting, for example, of a bellows with a whistle, the sound being amplified by a tiny horn. This latest book of prints for children thus becomes not only a delight for the eye but also for the ear.

A children's picture-book with animal noises [1898]

AN ELECTRIC RAILWAY FOR YOUNG FOLK

This year, 'les articles de Paris,' which come into the shops every year for the period of 'les étrennes,' the week of present-giving between Christmas and the New Year, include the midget railway of Monsieur Brillié. The young people, for whom it is intended, will no doubt be wide-eyed with surprise at seeing a train moving along a single rail, with engine and carriages suspended in mid-air. It is as well for boys and girls to know that this is a model of a new railway system that is now being tested in Great Britain and France: the 'monorail'!

The track consists of two copper strips kept parallel one above the other by insulat-

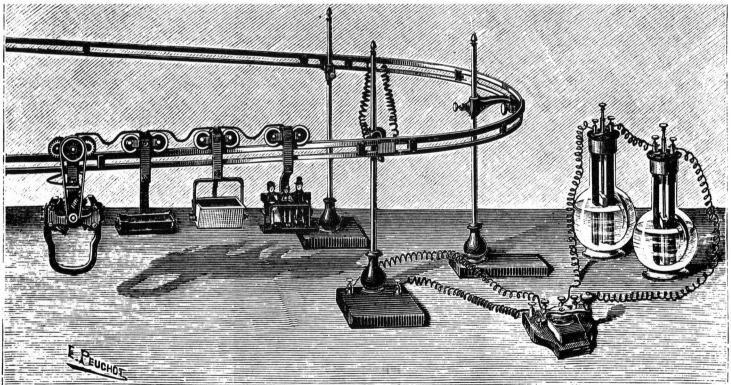

An electric railway for young folks [1887]

ing cleats and supported by poles. This makes it possible to introduce various bends and slopes into the track. The train is drawn by an engine containing an electric motor which receives its propulsive power from a series of potassium bichromate elements. The electric circuit includes a commutator with which the train can be driven forward, reversed or stopped. In addition, a variable resistance or rheostat is used to obtain various speeds.

A TOY DIRIGIBLE BALLOON

Whereas in former times men strove to use the resources of physics and mechanical engineering to create all manner of wondrous and mysterious machines for the amusement of the great, the spirit of our age is more inclined to present to young people as playthings objects which will help familiarise them with such mechanical devices. This is greatly frowned upon by many, but wrongly so, in our opinion.

No one will deny that ball games, bowling a hoop, whipping a top and other traditional games are more useful and healthy than playing and tinkering with mechanical toys because they provide the body with physical exercise. But there are circumstances under which such boisterous games cannot be played, and then mechanical toys which present the eye with rare surprises and keep little hands busily at work are certainly to be preferred to idleness and many other kinds of pastime.

It would be foolish to imagine that every young person who delights in playing with mechanical toys acquires so much knowledge about physics and mechanical engineering that he will become a specialist; it is only at a later age that he will inquire into the why and wherefore of it all. The main thing is the game itself, the production and perception of movement, and in this respect it is important that no wrong ideas should be imparted to the child. It may therefore rightly be required that the main principle of every mechanical plaything should be correctly

A toy dirigible balloon [1887]

185

What is seen by the spectators in a theatre [1883]

The scene as it is actually played [1883]

The haunted swing: left, illusion produced by a ride in the swing: right, true position of the swing [1894]

presented; and the divergences from this in secondary details can then be pointed out if need be by parents and those of superior knowledge. Considered from this viewpoint, the dirigible balloon of that practical aeronaut Mr Graham should not meet with disapproval.

His aim was to show in miniature, and in a playful way, how the rotary movement of a propeller can produce a forward motion. This he did by copying the method which is actually used to make balloons move by themselves through setting a propeller in motion. A miniature balloon does not have sufficient lifting-power to be able to carry the weight of any mechanism. Accordingly, only the shape of the real balloon with its attendant details could be considered for the purpose.

The balloon, made of papier maché, is of the familiar oblong shape, 19 inches long, and is suspended from a stand round which it can revolve like a chair-o-plane. This is counterbalanced by a small spherical balloon weighted with lead and suspended on the other side. The two-bladed propeller on the front of the balloon is connected through a system of cog-wheels to a shaft which is kept constantly in motion by means of tensed elastic bands wound up by means of a crank at the rear of the balloon. While the propeller is held fast, the crank is turned sixty times. The propeller is then released and, rotating rapidly, drives the balloon forward so that

it flies round at a fairly high speed for three minutes.

To add to the attraction, the rotating propeller acts upon a windlass in the gondola below the balloon by means of an endless cord. The arms of two dolls are worked by the drum of the windlass, thus creating the impression that these dolls are making the propeller move.

A somewhat more expensive version has a

musical box built into the base and a cord hangs from the gondola. When the balloon stops moving, the end of this cord points to one of the numbered squares painted on a printed plate below the machine so that it becomes a kind of game of chance. It would be a very spoiled young person indeed who would not find delight in playing with such an ingenious mechanical toy.

THE HAUNTED SWING
AN ODD SENSATION

No one who ever swung back and forth on a swing, increasing the arc, will have failed to wonder what would happen if it were to move so far as to make a complete circle. To satisfy their curiosity, they have only to travel to the city of Marseilles where a special type of swing was recently installed. Being mounted indoors, the swing accommodates some fifteen people who, when the swing appears to be in motion, undergo a curious illusion: although it is impossible for the swing to pass between the pivoting bar and the ceiling of the room, the occupants of the car seem to make several complete circles, until the swing is brought to a stop. The illusion is based upon the movements of the room itself. During the

'trip,' the swing is practically stationary whereas the room rotates about the suspending bar. At first, the swing may be given a slight push; the operators outside the room then begin to swing the room itself, which is really a large box journalled on the swing bar, starting it off to correspond with the movements of the swing, then making it go higher and higher until the swing itself appears to perform several complete rotations. At this time the people in the swing imagine that the room is stationary while they are whirling through space. Before the movements are brought to a stop, a number of back and forth swings are given to complete the illusion.

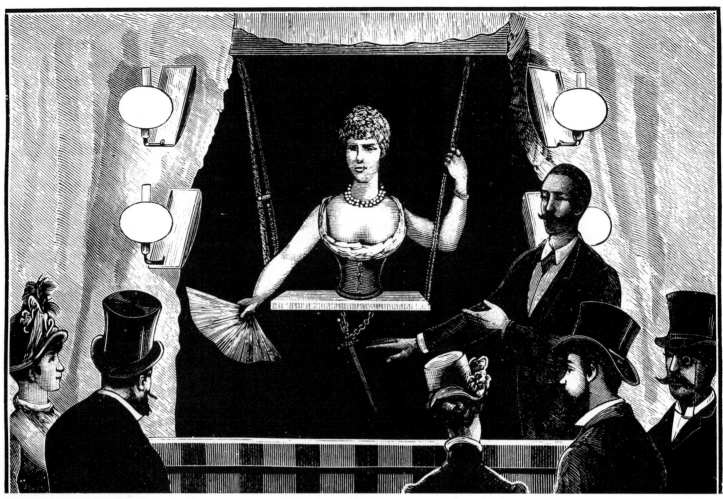

The woman without abdomen [1884]

Explanation of the phenomenon [1884]

ENGINEERING FOR PLEASURE-SEEKERS

In Paris, the latest advances in engineering science have been used for the entertainment of pleasure-seekers. This takes the form of a skilfully made sea serpent with a repulsive-looking head which wends its way through the Jardin d'Acclimatation or Zoological Gardens. An electric motor housed in the serpent's head and fed from a battery of accumulators provides the driving force. In spite of the uncanny appearance of the serpent, few can resist the temptation of making a trip on it.

An even more realistic sea monster is the midget battleship which is also propelled by electricity. With its single cannon, the crew—consisting of one admiral—can open fire on the coast; a bombardment full of smoke and bangs but without any risk to the 'target.'

The great sea serpent in the Zoological Gardens in Paris [1900]

Interior of the driver's cabin in the sea serpent, showing the electric motor and the accumulators [1900]

Miniature warship for entertainment park, firing harmless, smoke-producing shells [1899]

WALKING ON THE CEILING HEAD DOWN

A performance of considerable scientific interest has been produced in New York and other cities which is presented in the illustrations accompanying this article. In order to procure a perfect smooth surface to walk on, a board 24½ feet long is suspended from the ceiling, and near one end of this is a trapeze. The lower surface of the board is painted and is smooth and polished. The performer, who is known as Aimée, 'the human fly,' is equipped with pneumatic attachments to the soles of her shoes. Sitting in the trapeze with her face to the audience she draws herself upwards by the arms and raises her feet until they press against the board. They adhere by atmospheric pressure. She leaves the trapeze, and hangs head downwards, as shown. Taking very short steps, not over 8 inches in length, she gradually walks the length of the board backwards. She then slowly turns round, taking very short steps while turning, and

eventually returns, still walking backwards. This closes the performance.

To provide against accident a net is stretched under the board. The performer has frequently fallen, but so far no serious accident has happened. There is a certain art in managing to fall, as, if the shock were received directly by the spinal column, it might be very severe.

The attachment to the shoe is in general terms an India-rubber sucker with cup-shaped adhering surface. It is a disc 4½ inches in diameter and ⅝ inch thick. To its centre a stud is attached which is perforated near the end. This stud enters a socket fastened to the sole of the shoe. The socket is also perforated transversely. A pin is passed through the apertures, securing the hold between socket and disc. The socket is under the instep and is attached to the shank of the shoe sole.

A wire loop that extends forward under the toe of the shoe is pivoted on two studs which are secured on each end of the transverse central diameter of the disc. This loop is normally held away from the disc and pressing against the shoe sole by a spring. One end of the loop projects towards and over the rear edge of the disc. A short piece of string is secured to the India-rubber and passes through a hole in the extension of rearwardly projecting arm of the loop. The disc when pressed against a smooth surface is held fast by the pressure of the atmosphere. If now the loop is pressed towards the surface to which it adheres, the string will be drawn tight and will pull the edge of the India rubber away from the board. Air will rush in, and the adhesion will cease. As each new step is taken, one disc is made to adhere by pressure and the other is detached by the action described.

The power of the disc to sustain the weight of a performer may be easily calculated. Each sucker is $4\frac{1}{2}$ inches in diameter, and

What is seen by the spectators in the hall [1893]

Installation for producing the effect of equestrians riding the clouds in Richard Wagner's opera 'Valkyrie' at the Paris Théâtre de l'Opéra [1893]

191

contains therefore 16 square inches of surface. The full atmospheric pressure to the area would amount to 240 pounds. The stud and socket attachment provides a central bearing, so that the full advantage of this and of the disc is obtained, and a fairly perfect vacuum procured. As the performer only weighs about 125 pounds there is about 115 pounds to spare with a perfect vacuum.   [*1890*]

NEW SAFETY CURTAIN OF THE COMEDIE FRANCAISE IN PARIS

The frightful consequences of a fire in a theatre crammed with spectators have finally made us wake up to the fact that in building these places of public entertainment measures should be taken to prevent such disasters. Among these precautions is the possibility of separating the stage from the auditorium by means of an iron screen. In the Comédie Française in Paris they have gone a step further than this: the fire screen has been combined with the normal front curtain into an iron safety curtain which can be raised and lowered by means of an electric motor. This curtain first came into use in November 1892.

The curtain A weighs 880 pounds but is kept in balance by the counterweight D so that the motor F has only to contend with the friction. The curtain is carried by five cables 'a' which run along the pulleys 'o' and further over a drum B which can be rotated in either direction to pull the fire screen up or let it down. This movement is transmitted from the 2 horsepower motor by means of belts. The wires which supply the driving power to the motor terminate at an indicator-board in the prompter's box. By turning a crank in this direction or that, the prompter can cause the electric current to flow through the motor so that the curtain rises or falls. According to the effects desired on the stage, the curtain can be dropped at three different speeds and be raised at either of two speeds. The greatest speed at which it can be lowered is approximately $4\frac{3}{4}$ feet per second and the height over which it travels is $31\frac{1}{2}$ feet so that this fire curtain can be lowered within seven seconds.

Of course there is no lack of electric bells to give warning. There is one bell by the prompter's side which is brought into operation by means of two knobs in the wings. This tells the prompter when everything on the stage is ready and the curtain can go up. Furthermore, by moving the crank to bring the curtain down, the prompter sets various bells in motion which are located beside the head scene-shifter, and at various other places, to indicate the end of an act.

The theatre-going public are very much taken with the new fire curtain but the actors regret the disappearance of the old curtain which rose and fell in such slow but stately fashion.   [*1893*]

192